PROCESS SAFETY ENGINEER GUIDE

ALSO, BY P K SINGH

Books

CRACKING SAFETY AND HSE JOB INTERVIEW *
(Includes 200+ important Questions and Answers)

INTERVIEWER'S CHOICEST QUESTIONS
(Covers difficult Interview Questions and Answers

ACCIDENT & INCIDENT INVESTIAGATION *
(With Training Guide and Report Writing)

FIRE WATCH, SAFETY FOREMAN, RIGGER AND OPERATOR INTERVIEW GUIDE *

BEYOND SOLITUDE – Alone no More

CONVERSATIONS WITH CHILDREN
(The Missed Dialogues)

Please visit :

https://www.amazon.com/author/pksingh

PROCESS SAFETY ENGINEER GUIDE

Copyright © 2020 by P K Singh

The right of P K Singh to be identified as the Author of The Work has been asserted by him in accordance with The Copyright, Designs and Patents act 1988

1st Edition - Sep 2020
2nd Edition - July 2022
3rd Edition - Dec 2023

This is a complete guide for the Candidates going for Interviews, seeking a job as Process Safety Engineer.

All rights reserved. No part of this book may be reproduced or used in any manner without written permission of the copyright owner except for the use of quotations in a book review.

Book design by P K Singh

ASIN: B08HCXZXYR (e-book)
ASIN: B08HH9W9F9 (Paperback)
ASIN: B0B6L9GJKV (Hardcover)

ISBN-13: 979-8-6477-5434-9 (Paperback)
ISBN-13: 979-8-8407-2831-4 (Hardcover)

https://www.amazon.com/author/pksingh

Praise For

Process Safety Engineer Guide

A great guide !!!

The book explains what process safety management is, why it's important, and how to implement a PSM program with all of the 14 elements. It justly explains what you need to know to be fluent and ready for any Process Safety challenge. This is one of the best guides which ensure all Process Safety essentials are covered, from every angle.

The book will make you the dependable Process Safety domain authority by revealing the investigating strategy and taking steps to lower the Process Safety costs. Staff training and Management involvement needed have been described nicely keeping focus on understanding the process safety hazards associated with work environment.

The section on interview Question / Answers is great, I would highly recommend this book for job seekers and Professionals. This is one of the best guides which ensure all Process Safety essentials are covered, from every angle.

------ A.S., USA ----
(Amazon Verified purchase)

DEDICATION

To My Cherished Wife and Dearest Son & Daughter

This book is a testament to the love and support that each of you has brought into my life.

To my wife, your unwavering encouragement and understanding has been the cornerstone of my journey. Your love is the inspiration behind every word written here.

To my son, watching you grow into the incredible person you are fills my heart with pride

To my daughter, your grace, intelligence, and warmth bring joy to my days. May this book serve as a reminder of the boundless possibilities that await you and the love that surrounds you at every step of your journey.

Together, you form the tapestry of my life, and this book is a small reflection of the gratitude and love I feel for each of you. Thank you for being my pillars of strength, my sources of joy, and my greatest blessings.

With all my love,

P K Singh

TABLE OF CONTENTS

Dedication i
Table Of Contents iii
Acknowledgement v
Preface vii
1: INTRODUCTION 1
2: Interview Tips 11
3: Process Safety Engineer 14
4: 14 Elements of PSM 21
5: Hazard, Hazid, Hazan and Hazop 27
6: Process Hazard Assessments (PHA) 43
7: ASTM, BS & ASME Standard 49
8: Emergency Response Planning 55
9: Protective Actions for Life Safety 69
10: JHA/JSA (Job Hazard Analysis) 73
11: Risk Assessment 81
12: Hierarchy of Controls 97
13: Safe Work Method Statements 101
14: Layer of Protection Analysis (LOPA) 107
15: FEED 111
16: What is an EER diagram? 117
17: Safety Integrity Level (SIL) 127
18: HSE Audit 135

19: Leading & Lagging Indicators of Safety 149
20: Salary Negotiation 153
21: Interviewer's Ten Favourite Questions 157
22: Common Questions... 161
'Fire Triangle'... 163
23: General Safety Questions 179
24: Miscellaneous - Questions.. 187
25: Construction related Questions... 199
26: Simple Construction Site Safety Rules............................... 207
27: Pressure Relief Valve: Questions...................................... 211
28: Frequently asked Personal Questions............................... 217
29: Case Study (Confined space)... 233
List of Abbreviations 237
About the Author.. 239

ACKNOWLEDGEMENT

I would like to express my sincere gratitude and appreciation to all the people and organizations that have contributed so much to the creation of "Process Safety Engineering Guide."

I am very much thankful to almighty God for giving me beautiful surrounding, colleagues, good health and wisdom to write this Book for the benefit of worker's safety worldwide.

First, last and always, I wish to acknowledge the efforts of all Industry workers who despite the risks involved in the jobs carried out by them, execute high risk task, and wish that their Supervisors, Managers and Higher Management consider providing them safe working condition.

Second, I want to include in this group my former colleagues, whose privacy I wish to respect by not naming them here but whose contributions to this book are deeply grasped and highly appreciated.

Third, I have always been overwhelmed by my daughter Ms. Shibani's care and good wishes for me and my wife which is par above excellence by all standards. Her inspiration for me to write this book is highly commendable.

Fourth and finally, as a gratitude for all my life I am especially warmed by the thoughts of my helpmate, spouse, and partner Beena Singh, a woman of great wisdom, compassion and love, who has paved my way with highest thoughts about human relationships that do not need to remain fantasies but can be dreams come true.

PREFACE

(Navigating the Landscape of Safety)

Dear Readers,

Welcome to the prelude of "Exploring Safety in Industries: Your Essential Guide to Process Safety Engineering." As you hold this book in your hands, you're embarking on a journey into the heart of safety, where industries thrive, and people work with confidence. This preface serves as a compass, guiding you through the intentions and aspirations that inspired the creation of this comprehensive guide.

The Genesis of the Journey:

The idea behind this book emerged from a shared passion for safety and a realization that the world of Process Safety Engineering, while immensely significant, often remains shrouded in complexity. Safety should not be confined to the realm of experts; it should be a language spoken by all, accessible to anyone curious about the mechanisms that ensure our workplaces are secure and our industries flourish without compromise.

For the Curious Minds:

Whether you're a student taking your first steps into the professional world, a seasoned engineer seeking to expand your expertise, or someone simply fascinated by the orchestration of safety in industries, this book is designed for you. We believe that the journey into Process Safety Engineering should be as exciting as it is enlightening, as accessible as it is informative.

The Blueprint for Understanding:

In the initial chapters, we lay the foundation by demystifying the fundamentals of Process Safety Engineering. Picture it as a warm-up before a grand performance—a series of insights and explanations designed to make the seemingly complex world of safety comprehensible to all. We want you to feel comfortable navigating the landscape of safety, understanding its significance and embracing its principles.

Everyday Analogies:

To bridge the gap between theory and relatability, we've woven everyday analogies into the fabric of this guide. Whether you're baking cookies, planning a dinner party, or coaching a sports team, you'll find connections that make the abstract concepts of Process Safety Engineering more tangible. Safety, after all, is not an abstract notion; it's a practical, everyday necessity.

The Interview Trail:

For those on the cusp of entering the professional world or seeking new horizons in their careers, we've dedicated a section to the art of successful interviews. Imagine it as a mentor preparing you for a crucial conversation, arming you with the confidence and knowledge to navigate the intricate questions that often define career trajectories.

Advanced Chapters: A Gradual Ascent:

As we progress, we delve into more advanced concepts. Don't let the technical terms intimidate you; we unravel the intricacies of Hazard Identification (HAZID), Hazard and Operability Studies (HAZOP), and industry standards

PREFACE

(Navigating the Landscape of Safety)

Dear Readers,

Welcome to the prelude of "Exploring Safety in Industries: Your Essential Guide to Process Safety Engineering." As you hold this book in your hands, you're embarking on a journey into the heart of safety, where industries thrive, and people work with confidence. This preface serves as a compass, guiding you through the intentions and aspirations that inspired the creation of this comprehensive guide.

The Genesis of the Journey:

The idea behind this book emerged from a shared passion for safety and a realization that the world of Process Safety Engineering, while immensely significant, often remains shrouded in complexity. Safety should not be confined to the realm of experts; it should be a language spoken by all, accessible to anyone curious about the mechanisms that ensure our workplaces are secure and our industries flourish without compromise.

For the Curious Minds:

Whether you're a student taking your first steps into the professional world, a seasoned engineer seeking to expand your expertise, or someone simply fascinated by the orchestration of safety in industries, this book is designed for you. We believe that the journey into Process Safety Engineering should be as exciting as it is enlightening, as accessible as it is informative.

The Blueprint for Understanding:

In the initial chapters, we lay the foundation by demystifying the fundamentals of Process Safety Engineering. Picture it as a warm-up before a grand performance—a series of insights and explanations designed to make the seemingly complex world of safety comprehensible to all. We want you to feel comfortable navigating the landscape of safety, understanding its significance and embracing its principles.

Everyday Analogies:

To bridge the gap between theory and relatability, we've woven everyday analogies into the fabric of this guide. Whether you're baking cookies, planning a dinner party, or coaching a sports team, you'll find connections that make the abstract concepts of Process Safety Engineering more tangible. Safety, after all, is not an abstract notion; it's a practical, everyday necessity.

The Interview Trail:

For those on the cusp of entering the professional world or seeking new horizons in their careers, we've dedicated a section to the art of successful interviews. Imagine it as a mentor preparing you for a crucial conversation, arming you with the confidence and knowledge to navigate the intricate questions that often define career trajectories.

Advanced Chapters: A Gradual Ascent:

As we progress, we delve into more advanced concepts. Don't let the technical terms intimidate you; we unravel the intricacies of Hazard Identification (HAZID), Hazard and Operability Studies (HAZOP), and industry standards

Preface:

like ASME, ASTM, and BSI in a way that feels like embarking on an adventurous journey rather than navigating a complex maze.

Blueprints and Designs:

For those intrigued by the inception of industrial projects, we guide you through the Front-End Engineering Design (FEED) phase. We'll explore Safety Integrity Level (SIL), Layers of Protection Analysis (LOPA), and Design Emergency Escape Routes (EER)—terminologies that might sound formidable initially, but we'll break them down into stories that unfold like the pages of a captivating novel.

Learning from the Masters:

Every journey is enriched by the stories of those who've walked the path before us. In the section on Design Safety Case Studies, you'll encounter real-world narratives of successful projects—stories that serve as beacons of guidance, imparting lessons in problem-solving and decision-making in the realm of Process Safety Engineering.

Continuous Improvement: Audits and Indicators:

Our journey doesn't conclude with knowledge acquisition; it extends to the pursuit of excellence. In the chapters on HSE Audits and Health & Safety Performance Indicators, we introduce tools that facilitate continuous improvement. Think of them as instruments for refining the symphony of safety, ensuring that every note played contributes to a safer, more harmonious workplace.

A Guiding Light:

"Exploring Safety in Industries" is more than just a book; it's a guiding light. It's an invitation to understand, appreciate, and actively participate in the world of Process Safety Engineering. Safety is not a constraint; it's an enabler, a catalyst that allows industries to flourish and individuals to thrive in their workplaces.

Your Journey Begins Here:

As you embark on this journey, embrace the spirit of curiosity and the joy of learning. The world of safety is not reserved for experts alone; it's a shared responsibility that each of us can contribute to and benefit from. Let this guide be your companion, your mentor, and your source of empowerment in the realm of Process Safety Engineering.

To Safety, Knowledge, and Beyond:

We extend our deepest gratitude to you, the reader, for choosing to explore safety with us. May this guide inspire you, empower you, and kindle a newfound appreciation for the intricate dance of safety in industries. Your journey into Process Safety Engineering starts here, and we're honoured to accompany you every step of the way.

With warm regards,

P K Singh

Chapter 1

: INTRODUCTION

Introducing "Process Safety Engineer Guide," a fascinating handbook for anybody curious about the art and science of maintaining workplace and industrial safety. This book is your ticket to the world of process safety engineering, whether you're an inquisitive student, a recent graduate starting your career, or a seasoned professional trying to expand your knowledge.

Setting the Stage:

Envision a bustling manufacturing facility with occupied workers and humming machines. Amidst this orderly chaos, Process Safety Engineering is the unsung hero at work. Our book aims to make the complex subject of industry safety exciting and understandable for all readers by uncovering the secrets of this unsung hero.

Why Process Safety Matters:

Ever wonder how some industries provide goods and services without endangering people or causing accidents? That's where Process Safety Engineering comes in. It's similar to having a superhero ensuring that everything runs well and, most importantly, that no one is harmed.

The Journey Begins:

In the first few chapters, we'll take you on a trip to understand the basics of Process Safety Engineering. Think of it as learning your ABCs before you can read and write. We'll examine the tools and research used the significance of safety, and the rationale for industry adherence to particular standards and laws. It serves as the cornerstone and framework that establishes the prerequisites for a safe and effective workplace.

Unlocking the Secrets:

Have you ever wondered how businesses make sure their operations are both safe and productive? Our article delves into the toolkit of Process Safety Engineers, revealing techniques that maintain seamless operations. We'll walk you through the complex process of making workplaces secure, from seeing possible threats to determining how to handle them.

Bringing it to Life:

We'll compare things to real-world situations to make them more approachable. Assume you are in your kitchen, making cookies. You would observe safety precautions, such as wearing oven mitts, and double-check the recipe to make sure you have all the ingredients. Process safety engineers also have a "recipe" for making sure that industries are safe. The principles become more concrete and easier to

1: INTRODUCTION:

understand when they are connected to everyday behaviours and workplace safety.

The Heart of the Matter:

Key ideas such as Hazard Identification (HAZID) and Hazard and Operability Studies (HAZOP) are fundamental to Process Safety Engineering. Don't be alarmed by the large words! We'll break these ideas down into manageable chunks and show you how they function in the real world with anecdotes and examples.

Standards and Guidelines Made Simple:

Ever ponder why different sectors adhere to particular regulations and standards? We'll simplify technical jargon and provide an explanation of standards like as ASME, ASTM, and BSI that will make them seem as simple as deciphering a code. You'll see why these guidelines are equivalent to a reliable manual, guaranteeing that everyone follows the same principles in order to preserve safety.

Emergency Plans and HSE Procedures:

Let's now pretend that you are throwing a dinner party. You would budget for unforeseen events, such as running out of materials or someone spilling a drink. In a similar vein, industries maintain Health, Safety, and Environment (HSE) protocols and emergency plans in case of unanticipated events. We'll walk you through developing efficient emergency plans so that in the industrial environment, safety becomes second nature.

Interview Success:

We have solutions for anyone wishing to advance in their careers or enter the workforce. A wealth of frequently asked interview questions for roles in Process Safety Engineering is included in the book. Every question serves as a stepping stone to assist you move confidently and gracefully through interviews.

Advanced Concepts in Layman's Terms:

Advanced ideas like Job Hazard Analysis (JHA), Risk Assessment, and Safe Work Method Statements (SWMS) will be covered as we go farther. Consider these as advanced chapters in a gripping book where every page offers a fresh perspective. Your guides will be real-world examples and useful applications, which will help the complex, seem approachable and familiar.

From Design to Implementation:

Ever ponder the beginning of a significant construction project? We will walk you through the Front-End Engineering Design (FEED) stage, which lays out the winning strategy. Although phrases like Safety Integrity Level (SIL), Layers of Protection Analysis (LOPA), and Design Emergency Escape Routes (EER) may seem intimidating, we'll break them down into approachable scenarios and captivating stories.

Learning from Experience:

The book includes Design Safety Case Studies, which are true accounts of accomplished projects that offer insightful lessons. It's similar like sitting down to hear from seasoned pros about their experiences. These case studies will serve as your mentors, offering you insights into the realm of Process Safety Engineering decision-making and problem-solving.

1: INTRODUCTION:

Quality Assurance and Performance Metrics:

But that's not where our trip ends. We will discuss tools that guarantee continual development, such as Health and Safety Performance Indicators and HSE Audits. These are your tactics for improving the team with every game, if you were a coach overseeing a team. HSE audits and performance indicators are the instruments used in the industrial context to attain superior safety standards.

A Transformative Learning Experience:

"Process Safety Engineer Guide" is an immersive learning experience as well as a book. We cordially encourage you to set off on a journey that will revolutionize the way you think about industry safety. This guide is designed to meet you where you are and take you to new heights in the field of Process Safety Engineering, regardless of your experience level.

This book has been prepared in an easy-to-understand format, especially made for the Process Safety Engineers and job seekers in this field.

The book gives detailed information Process Safety requirements of all industries including Chemical, Construction, Petrochemical and Oil & Gas Industry.

A comprehensive section on typical Question and Answers generally asked in the interviews has been consolidated and presented in an easy-to-understand format. The topics covered are exhaustive and meet the industry needs. They are classified into: -

Basic Theory of Process Safety

Interview preparation tips and techniques

Guidance to crack the interview

Effective communication skills

Presentation skills

This Book is:

"............ An easy-to-understand compilation on interview questions for candidates"

"............ a complete interview guide

"............ one of the best professional books on the subject"

"............ the questions and answers are taken from actual interviews conducted by Clients"

.... a must if you are looking for a Process Safety job"

"..... the book outlines how to turn a Job Interview into a Job Offer"

Do you have answers for the following tough Questions (if not then this Book will help you in replying when you are caught in such questions)? -

1. Why should we hire you as a Process Safety Engineer?

2. Tell me about yourself?

3. What are your biggest weaknesses?

4. What are your biggest strengths?

5. What is your ideal work environment?

6. Why do you want this, Job?

1: INTRODUCTION:

7. Where Do You See Yourself in Five Years?

8. Tell me about your dream job

9. Why are you leaving your current position?

10. What makes you different from other Applicants?

11. How do you handle disagreements with your Boss?

12. What motivates you?

13. What are the biggest challenges you feel are faced by the industry?

14. What do you hope to accomplish in this position?

15. How do you deal with pressure?

16. What are your expectations for this position?

17. Would you like to ask us anything?

The Interview tips have been written in clear and concise manner. Generally requested Job descriptions are consolidated from various companies and put together.

Interviewer's choicest questions like "tell me about yourself etc. is explained in detail with answers on how to answer them. Also, an exhaustive Question and answer guide for frequently asked questions have been provided.

General Information

For A "Successful" Interview

Chapter 2:

INTERVIEW TIPS

The following information has been compiled on the basis of actual interviews conducted by companies.

The inputs of candidates hired in recent times are also incorporated.

The tips are exhaustive and cover almost everything for making a positive impression.

Entering the Interview Room

The interviewer should be greeted by "Good morning, Good afternoon or Good evening". Alternately wishing "Good day" at any time of the day is also great.

Have a firm handshake with the interviewer/s with a smile on your face. Address the interviewer by his / her first name (if you know) and talk/listen while looking into his eyes.

Do not sit until asked to do so.

Prepare yourself before going for the interview; try to get as much information about the organization and the job position as possible. Visit the website of the Company, read about them, get information about the job position and the salary range offered etc. If the job is through any agency, then they can give you more information on this.

Read your own Bio-Data Resume thoroughly. If someone else made your Resume you have to be more cautious on the details mentioned. You must know the dates of your previous jobs, company wise in ascending order of dates i.e. from older date/year to latest.

Reach the interview venue before scheduled time. You should know the exact location of the office, and how you can reach there. It is advisable to reach at least 10 to 15 minutes early, so you can get adjusted.

Drink water or have a cup of tea so that you do not feel thirsty soon.

Dress in a smart way; don't put colourful dress with eye catching designs. It is best to put on formal dress.

Be decent, look confident & successful.

When you are asked to say something about yourself, DO NOT talk about non-work or personal life. Most candidates start talking about their family and hobby way too much. No one is interested in it, except you. Talk about your work & your duties.

Do not speak negative about your past organization.

At the end of the interview, if the interviewer asks you – "Do you have any questions" Be enthusiastic and ask

2: Interview Tips:

some questions about the company, its goals, about the job position etc.

Do not forget to thank the interviewer for the time they took out for you.

If it was a Skype interview of you and you have the email id of HR department of the company then you may send a "Thank You Letter" after the interview, particularly if you know the interview went well and you are interested in taking up that position. This will refresh the interviewer's memory and keep you on top of the list. If you do not receive a response a week later, go ahead and call.

In case a recruiting agency was involved in the interview process, then let them know whether you are interested or not. Many times, the organization is waiting for an answer from the candidate, before they say yes.

In short there is Need to Focus on: -

1. Appearance.
2. Enthusiasm.
3. Strong Handshake.
4. Punctuality.
5. Ask Questions.
6. Confident Answers.
7. Be Polite & Soft-spoken.
8. Be Courteous.
9. Show Maturity.
10. Present High Moral Standards.

Chapter: 3

PROCESS SAFETY ENGINEER

Who is a process safety engineer?

A Process Safety Engineer is a professional that is in charge of assuring the safety of industrial processes and facilities during their design, operation, and maintenance. Their main goal is to prevent accidents,

3: Process Safety Engineer:

mishaps, and the release of hazardous materials that could endanger people, the environment, or property. These engineers operate in a variety of industries, including chemical production, oil and gas, pharmaceuticals, and others that involve sophisticated procedures and potentially hazardous materials.

The Process Safety engineers are employees responsible for developing the risk assessment and designing safe operating practices for processing and manufacturing companies. They provide technical leadership and support to identify hazards, assess risks and provide cost-efficient management solutions.

Since the potential for disaster is huge when hazardous chemicals are stored, it is the responsibility of Process Safety Engineer in industries that use them to ensure that dangerous materials are properly controlled.

Key Responsibilities:

1. **Risk Assessment:**

    Process Safety Engineers do extensive risk assessments to detect potential hazards and estimate the level of risk associated with diverse industrial processes. To systematically examine potential hazards, they employ methodologies such as Hazard and Operability Studies (HAZOP), Fault Tree Analysis (FTA), and Failure Modes and Effects Analysis (FMEA).

2. **Safety System Design:**

    They design and execute safety measures and safeguards to prevent and lessen the effects of probable accidents. This comprises the design and installation of safety instrumentation, emergency shut-down systems, and other safety measures.

3. **Compliance with Standards:**

   Process Safety Engineers make sure that industrial processes adhere to all applicable safety standards, laws, and regulations. Standards established by organizations such as the American Society of Mechanical Engineers (ASME), the American Society for Testing and Materials (ASTM), and the British Standards Institution (BSI) may fall into this category.

4. **Emergency Response Planning:**

   They create and carry out emergency reaction strategies in the event of an unforeseen crisis. Coordination with other departments, training staff, and drills are all part of ensuring a quick and successful reaction in the event of an emergency.

5. **Continuous Improvement:**

   Process Safety Engineers work on continuous improvement by examining and updating safety procedures and systems on a regular basis to address evolving risks and industry best practices.

6. **Training and Education:**

   They train employees on safety protocols and procedures to ensure that everyone involved in the process understands and follows safety requirements. This could involve holding safety courses, seminars, and awareness campaigns.

3: Process Safety Engineer:

Education and Skills:

A bachelor's degree or higher in chemical engineering, process engineering, or a related discipline is normally required for Process Safety Engineers. They must be well-versed in chemical processes, thermodynamics, and fluid dynamics. Effective communication and problem-solving abilities, in addition to technical knowledge, are required for success in this profession. It is also necessary to be able to work cooperatively with cross-functional teams and adapt to changing scenarios.

Certifications:

To demonstrate competence and commitment to the profession, certifications such as the Process Safety Professional (PSM) or Certified Process Safety Engineer (CPSE) may be pursued. These credentials frequently necessitate a mix of schooling, work experience, and passing a test.

A TYPICAL JD OF- PROCESS SAFETY ENGINEER

Job Description (JD):

Working with the technical/operations teams on the Risk Management process (i.e., HAZID, Risk Assessment and HAZOP, etc.).

Providing guidance on all process safety standards, codes & guidelines in projects and operations (ASME, ASTM, BSI etc.).

Write and assist subsidiaries to write HSE procedures and emergency plans, identifying potential emergency scenarios and developing emergency preparedness plans.

Reviewing all new project contractors' relevant HSE documents e.g. HSE-MS; project HSE plans, JHA, Risk assessment, safe work method statement etc.

Facilitating and participating in technical process safety studies e.g. FEED Reviews, HAZOP, HAZIDS, SIL, LOPA, Design EER, Design safety case studies etc.

Maintaining a database of all HSE technical safety studies.

Planning and conducting HSE audits and reviews toward project team and subcontracted activities.

Developing & monitoring Health & Safety Performance Indicators for management review.

Plan, manage, coordinate and monitor the Safety, Health, Environment and Security issues pertaining to the project deliverables comply with approved project plans.

Minimum Qualification / Requirements

Bachelor Degree (preferable in Science Stream)

NEBOSH IGC is mandatory

Possession of valid trainings

Knowledge and/or Experience

Total 3 to 10 years varied Industrial experience in safety.

Specific experiences in managing HSE aspects & contracts and have good understanding of safe work practices.

Knowledge and proficiency in HSE Regulations and Best practices (OSHA, ISO 14001, OHSAS 18001).

Technical and Business Skills

Proficient in written and spoken English.

3: Process Safety Engineer:

Computer literacy, including the use of spread sheets, databases.

Ability to communicate effectively with internal/ external clients.

Good adaptability to multinational environment, with wide exposure to various cultures and customs.

Minimum Medical requirements: -

Medical requirements are based on individual companies' policies, tests are conducted at Company authorized Clinics in various cities.

Desirable Qualifications/Experience

These will be an added advantage but are not mandatory.

Trained in Basic First Aid techniques from International accredited courses like Medic First Aid.

Good knowledge of lifting and rigging techniques.

Competency in inspection of advance Marine fire fighting systems.

Experience of Installation Barges, topside installations, DP Diving support vessels on Air diving and Sat diving, Pipe Laying Barges in any operational role.

Experience on Offshore production Platform Hook Up and Commissioning in any operational role.

Competent in inspection of emergency marine floatation devices like Life Jackets, Life Buoys, Work Vests.

Define criterion for work scope like Cascade system for Marine vessels, Jack up Barges and periodically inspect the cascade system.

Trouble shoot the cascade system installed on the Jack up Barges and Marine Vessels. Deliver refresher Tool Box Talks regarding use of BA systems, Cascade systems.

Inspect the Life Boat and deliver refresher Tool Box Talks regarding launch and use Life Raft to construction crew.

Possession of National Seaman Book or "Seaman's Seagoing Service Record Book" and where applicable copies of the main page and the pages with vessel entries.

Knowledge about the use of reference to Tide charts to plan marine activities in Splash zone and advice on planning of different activities involving current and stability of marine vessel with reference to Weather criterion defined in Operations Execution Plan or Project HSE Plan as applicable.

Radiation Process Safety Engineer Training.

Explosive Handling Safety.

Knowledge about Vessel-to-Vessel Protocols, Knowledge about MARPOL and other such legislations.

Conclusion:

In short, a Process Safety Engineer is critical to the safety of industrial processes and the well-being of persons and the environment. Their work is critical to balancing productivity and safety, preventing accidents, and constantly upgrading safety measures in complicated industrial environments.

Chapter: 4

14 ELEMENTS OF PSM

The 14 Elements of Process Safety Management are as under: -

1. Employee participation

Description: Perhaps one of the most important mandates, the employee participation clause requires that employees—including production and maintenance staff—be involved in every aspect of the PSM programs at their respective worksites. They must also be represented at the meetings where PSM-related issues are discussed. OSHA requires employee participation to be documented.

Importance: Engaged and informed employees give vital ideas, establishing a culture of safety and continual development.

2. Process safety information (PSI)

Description: As per OSHA, "The employer shall complete a compilation of written process safety information before conducting any process safety hazard analysis required by the standard."

Hence, all workers should be able to access and understand the technical data regarding the related risks they could face on the job. They should be able to compile and maintain comprehensive information about the chemicals, technology, and equipment used in the process.

Importance: Access to accurate and up-to-date information is critical for assessing and managing potential hazards.

3. Process hazard analysis (PHA)

Description: Process Hazard Analysis, the core technical component of PSM, requires engineers and maintenance leaders to analyse the consequences of safety failures. These assessments should be carried out in groups, and OSHA mandates that each group include at least one person who is "knowledgeable in the specific process hazard methodology being used."

Importance: PHAs aid in the identification of potential dangers and the execution of safety measures to mitigate those risks.

4. Operating procedures

Description: Developing and maintaining documented procedures for properly carrying out process tasks.

4: 14 Elements of PSM:

During Turnarounds and emergency shutdowns there is wide variety of potential chemical hazards. OSHA guidelines want to ensure that companies have plans for keeping everyone safe as they start back up.

Importance: Operating procedures that are clear and standardized help to prevent deviations that could lead to accidents.

5. Training

Description: Workers who perform processes involving hazardous chemicals must be trained. Their training should have been provided by a qualified source, whether first-party or third-party. OSHA requires that all training be documented.

Importance: Well-trained workers are better suited to handle processes safely and effectively in the event of an emergency.

6. Contractors

Description: Regular employees and contractors must be made aware of any hazards that may arise during the course of their duties. It is the responsibility of the employer to notify the contract employers of any known potential fire, explosion, or toxic release hazards associated with the contractor's job and the process.

Importance: To create a united safety culture, contractors operating on-site should adhere to the same safety requirements as regular employees.

7. Pre-start-up safety review

Description: OSHA expects businesses to conduct pre-start-up safety reviews for both new and modified facilities. It also applies if the procedural modifications affect only a single component or process.

Importance: Assists in identifying and correcting any safety concerns before they pose a risk during normal operations.

8. Mechanical integrity

Description: Conducting a safety review prior to the start-up of a new or modified procedure.

Pressure vessels, Storage tanks, Piping systems and Ventilation systems need periodic and documented inspections.

The testing procedures must be recognized and accepted as good engineering practice.

Importance: Aids in the identification and correction of potential safety concerns before they represent a risk during normal operations.

9. Hot work permit

Description: Implementing a permit mechanism for actions such as welding or cutting that could introduce an ignition source.

Every employer needs to issue permits to employees and contractors who weld or perform other high-temperature work near covered processes. They also need to train their personnel to post and file these permits when necessary.

Importance: Ensures that fire and explosive threats are avoided during hot work operations.

10. Management of change (MOC)

Description: There should be standard procedures for managing changes to processes, equipment and procedures.

The following points need to be looked into: -

- Basis for change.

4: 14 Elements of PSM:

- Impact of the change on worker safety and health.
- Necessary modifications to operating procedures.
- Time period for the change.
- Authorizations for the proposed changes.

Importance: Contributes to the prevention of unintended effects by assessing and regulating the potential impact of modifications on process safety.

11. Incident investigation

Description: Conducting extensive investigations of incidents in order to find root reasons and prevent recurrence. All incidents which could have resulted in an accident should be investigated.

Importance: Learning from incidents is critical for on-going progress and avoiding similar incidents in the future.

12. Emergency planning and response

Employers must create emergency plans for handling emergency situations.

13. Compliance audits

Description: Employers must get Compliance audits every three years to verify that the procedures and practices developed under the standard are adequate and are being followed. It is also required to retain at least their two most recent audit reports.

Importance: Audits provide an independent evaluation of the system's performance and identify areas for improvement.

14. Trade secrets

Description: Ensuring that staffs have access to required safety information while protecting proprietary process information.

Importance: Balancing the need for safety information transparency with the security of proprietary information.

Chapter 5

HAZARD, HAZID, HAZAN AND HAZOP

(Risk Management Techniques)

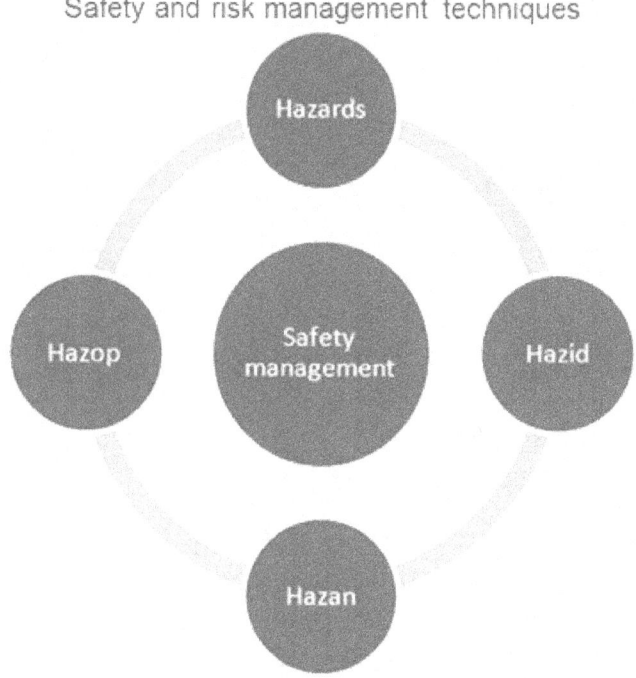

The below mentioned studies are important safety and risk management techniques used by Process Safety Engineers: -

1. Hazard
2. Hazid (hazard identification)
3. Hazan (hazard analysis)
4. Hazop (hazard and operability studies)

Hazard

Hazard is a condition, event, or a circumstance that can lead to an unplanned or undesirable event. It has the potential to cause serious harm to the individual or the environment. Any process or situation that has the potential to cause harm constitutes a hazard.

Every hazard will not result into a disaster, but every disaster is the result of a hazardous condition.

The following are the key indicators in terms of the situations arising.

- Magnitude
- Intensity
- Weather
- Time
- Duration
- Frequency
- Prediction possibility
- Related hazards
- Cascading effects

The outcome of the above helps in estimating the relative hazard posed by any event. Hazard depends on the magnitude and duration of any event. Examples of catastrophic intensity include events like flood, hurricane etc.

5: Hazard, Hazid, Hazan and Hazop:

In some cases, minor events may lead to catastrophic failures. Events that tend to trigger other events or that can cause cascading effects are higher in their hazard potential.

Hazard identification or Hazid

Hazid stand for hazard Identification. All industries with a high risk often require that all hazards with the potential to cause a major accident are identified.

A hazard identification study (HAZID) is the procedure used for hazard identification in workplace health and safety. It is conducted by breaking down all processes in a worksite for detailed analysis. Once a hazard is identified, the hazard identification study reviews the effectiveness of the selected measures and expands the safety measures when required to reach a tolerable residual risk.

Hazid requires high level hazard identification technique which is commonly applied on an area by area basis to hazardous installations. It is the systematic method of identifying hazards to prevent and reduce any adverse impact that could cause injury to personnel, damage or loss of property, environment and output, or become a liability. It is a component of the risk assessment and risk management and is being used to determine the adverse effects of exposure to hazards and to plan necessary actions to mitigate such risks.

It provides a structured approach to identify hazards, potential undesirable consequences, and evaluate the severity and likelihood of what is identified. Hazid involves machine or equipment designers, management and end users, and ensures a full identification of hazards and safeguard procedures in a workplace.

The two possible purposes in identifying hazards are:

To obtain a list of hazards for subsequent evaluation using other risk assessment techniques. This is also known as 'failure case selection'.

To perform a qualitative evaluation of the significance of the hazards and the measures for reducing the risks from them. This is also known as 'hazard assessment'.

Through Hazid, management identifies, in consultation with employees, contractors and safety personnel the following: -

All reasonably foreseeable hazards at the plant that may cause a major accident

The kinds of major accidents that may occur at the plant, the likelihood of a major accident occurring and the likely consequences of a major accident.

Hazid should be an on-going process to ensure existing hazards are known, and new hazards are recognized before they are introduced due to the following: -

- Prior to modification of any facility
- Prior to change in the method of work
- Before and during abnormal operations and during troubleshooting
- Plant condition monitoring, early warning signals, employee feedback from routine participation in work, and after an incident

Hazid is critical to the safety of the plant, equipment, and operating personnel. The benefits of Hazid studies include the following: -

- It is a flexible method which is applicable to any type of installation
- It reveals hazards at an early stage before they happen

5: Hazard, Hazid, Hazan and Hazop:

- It leverages the experience of operating personnel as part of the team
- It identifies hazards, cause and consequences as well as preventive measures
- Hazards are recorded and managed to be avoided, mitigated or highlighted
- It establishes screening criteria for hazards
- Non critical hazards are documented to demonstrate that the events in question could be safely ignored

Hazard analysis or Hazan

Hazan (Hazard analysis) is a term used in Process Safety Engineering for the logical, examination of an item, process, condition, facility, or system to identify and analyse the source, causes, and consequences of potential or real unexpected events which can occur.

A hazard analysis considers system state (e.g. operating environment) as well as failures or malfunctions. Hazan is the identification of undesired events that lead to a hazard. It is the analysis of the mechanisms by which these undesired events could occur, by the estimation of the consequences. Every hazard analysis consists of the following three steps: -

1. Estimating how often the incident will occur.
2. Estimating the consequences for the employees, the process, the plant, the public and the environment.
3. Comparing the results of first two steps with a target or criterion to decide whether or not action to reduce the probability of occurrence or to minimize the consequences is desirable, or whether the hazard can be ignored.

Hazan is the first step in the process used for the assessment of the risk. The result of a hazard analysis is the identification of different type of hazards. A hazard is a potential condition which either exists or does not exists (probability is 1 or 0). It may be in single existence or in combination with other hazards (called events) and conditions become accident (mishap). This scenario has a probability (between 1 and 0) of occurrence. Often a system has many potential failure scenarios. Risk is the combination of probability and severity. Preliminary risk levels can be provided in the hazard analysis. The main goal of Hazan is to provide the best selection of means of controlling or eliminating the risk.

Severity definitions used in Hazard analysis

Tab (i) Safety related severity definitions

Severity	Definition
Catastrophic	Results in multiple fatalities and/or loss of the system
Hazardous	Reduces the capability of the system or the operator ability to cope with adverse conditions to the extent that there would be: (i) Large reduction in safety

5: Hazard, Hazid, Hazan and Hazop:

margin or functional capability

(ii) Operators physical distress/ excessive workload such that operators cannot be relied upon to perform required tasks accurately or completely

(iii) Serious or fatal injury to the work men

(iv) Fatal injury to personnel and/or general public

Major	Reduces the capability of the system or the operators to cope with adverse operating conditions to the extent that there would be: (i) Significant reduction in safety margin or functional capability (ii) Significant increase in operator

workload

(iii) Conditions impairing operator efficiency or creating significant discomfort

(iv) Physical distress to workmen including injuries

(v) Major occupational illness and/or major environmental damage, and/or major property damage

Minor | Does not significantly reduce system safety. Actions required by operators are well within their capabilities. Include:

(i) Slight reduction in safety margin or functional capabilities

5: Hazard, Hazid, Hazan and Hazop:

(ii) Slight increase in workload

(iii) Some physical discomfort to the operators

(iv) Minor occupational illness and/or minor environmental damage, and/or minor property damage

No safety effect	Has no effect on safety	

Tab (ii) Likelihood of the occurrence of an event

Likelihood	Definition
Probable	Qualitative – Anticipated to occur one or more times during the entire system/operational life of an equipment

	Quantitative – Probability of occurrence per operational hour is greater than 0.00001
Remote	Qualitative – Unlikely to occur to each item during its total life. May occur several times in the life of an entire system or group of equipment.
	Quantitative – Probability of occurrence per operational hour is less than 0.00001, but greater than 0.0000001
Extremely Remote	Qualitative – Not anticipated to occur to each item during its total life. May occur a few times in the life of an entire system or group of equipment.
	Quantitative: Probability of occurrence per operational hour is less than 0.0000001, but greater

5: Hazard, Hazid, Hazan and Hazop:

	than 0.000000001
Extremely Improbable	Qualitative – So unlikely that it is not anticipated to occur during the entire operational life of an entire system or group of equipment
	Quantitative – Probability of occurrence per operational hour is less than 0.000000001

Some of the terms used in the Hazan and their definition are given below.

- Hazard rate – The rate (occasions/year) at which hazards occur.
- Protective system – A device installed to prevent the hazard
- Test interval – The time interval between the testing of a protective system, and its replacement if necessary.
- Demand rate – The rate (occasions/year) at which a protective system is called to act.
- Failure rate – The rate (occasions/year) at which a protective system develops faults (fail-danger/fail-safe).

HAZOP

Hazard and operability study or Hazop

A hazard and operability (Hazop) study is a technique used for hazard identification, and for the identification of design deficiencies which may give rise to operating problems. It identifies and evaluates problems that may represent risks to personnel or equipment, or prevent efficient operation. Hazop is most commonly applied to systems which transfer or process hazardous substances, or activities where the operations involved can be hazardous and the consequences of failure to control hazards may be significant in terms of damage to life, the environment or property.

A Hazop study is carried out by an experienced multi-discipline team, facilitated by a Hazop leader. The Hazop technique is qualitative, and aims to stimulate the imagination of participants to identify potential hazards and operation issues. The Hazop technique was initially developed to analyse chemical process systems and mining operation process but has later been extended to other types of systems and also to complex operations.

Hazop is based on a theory that assumes risk events are caused by deviations from design or operating intentions.

As a risk assessment tool, Hazop is often described as the following:

- o A brainstorming technique.
- o A qualitative risk assessment tool.
- o An inductive risk assessment tool, meaning that it is a 'bottom-up' risk identification approach, where success relies on the ability of subject matter experts to predict deviations based on past experiences and general subject matter expertise.

5: Hazard, Hazid, Hazan and Hazop:

The objectives of the Hazop studies are the following:

- Identify hazards and operation issues associated with the design.
- Identify deviations from design intent, deviation causes, consequences, and safeguards.
- Provide an action list with due dates and identify appropriate person/discipline to progress the action to close out harm.

Hazop is best suited for assessing hazards in facilities, equipment, and processes and is capable of assessing systems from multiple perspectives which include the following.

- Design – This includes (i) assessment of system design capability to meet user specifications and safety standards, and (ii) identification of the weaknesses in systems.

- Physical and operational environments – It includes assessment of the environment to ensure system is appropriately situated, supported, serviced, contained, etc.

- Operational and procedural controls – These includes (i) assessing engineered controls (ex: automation), sequences of operations, procedural controls (ex: human interactions) etc., and (ii) assessing different operational modes such as start-up, standby, normal operation, steady and unsteady states, normal shutdown, emergency shutdown, etc.

Advantages of Hazop technique are as follows.
- It is helpful in confronting hazards that are difficult to quantify
- Helps in analysing hazards rooted in human performance and behaviours
- Hazards that are difficult to detect, analyse, isolate, count, predict, etc.
- Methodology doesn't force any one to explicitly rate or measure deviation probability of occurrence, severity of impact, or ability to detect

- Built-in brainstorming methodology

- Systematic and comprehensive methodology

- More simple and intuitive than other commonly used risk management tools

Disadvantages of Hazop technique are as follows.
- No means to assess hazards involving interactions between different parts of a system or process

- No risk ranking or prioritization capability

- Teams may optionally build-in such capability as required

5: Hazard, Hazid, Hazan and Hazop:

- No means to assess effectiveness of existing or proposed controls (safeguards)

- May need to interface Hazop with other risk management tools

Chapter 6

PROCESS HAZARD ASSESSMENTS (PHA)

Process Hazard Assessment (PHA) is a systematic assessment of the potential hazards associated with an industrial process. A PHA can provide information intended to make decisions for improving safety and reducing the consequences of unwanted or unplanned releases of hazardous chemicals.

A PHA analyses potential causes and consequences of undesirable events and it focuses on equipment, instrumentation, utilities, human actions, and external factors that might impact the process.

A multidisciplinary team looks at several parts of the process throughout a PHA, taking into account things like tools, chemicals, protocols, and outside influences. Hazard and Operability Studies (HAZOP), Failure Modes and Effects Analysis (FMEA), and What-If Analysis are common techniques used in PHAs. The group evaluates possible departures from standard procedures, ascertaining the probability and impact of such deviations.

A PHA's main objectives are to identify possible risks, evaluate the risks that go along with them, and suggest actions to reduce or eliminate those risks. Organizations may increase safety protocols, plan for emergencies better, and adopt preventive actions by methodically assessing process components and their interdependencies.

PHAs are essential parts of process safety management systems, which help to prevent catastrophic events and create safer working environments in a variety of industries, including as the manufacturing, petrochemical, and chemical sectors.

The PHA methods are qualitative in nature the method selected depends on the complexity of the process, the length of time a process has been in operation and whether a PHA has been conducted on the process before, and if the process is unique, or common to the industry.

Methodology of PHA

There are a variety of methodologies that can be used to conduct a PHA, including but not limited to:

– Checklist;

– What if?

– Hazard and Operability Study (HAZOP);

– Hazard Identification (HAZID);

– Failure Mode and Effects Analysis (FMEA).

6: Process Hazard Assessments (PHA):

FSES Typical HAZOP Workflow

The following workflow provides a typical guide on structuring the hazard assessment or quantitative risk assessment workshop.

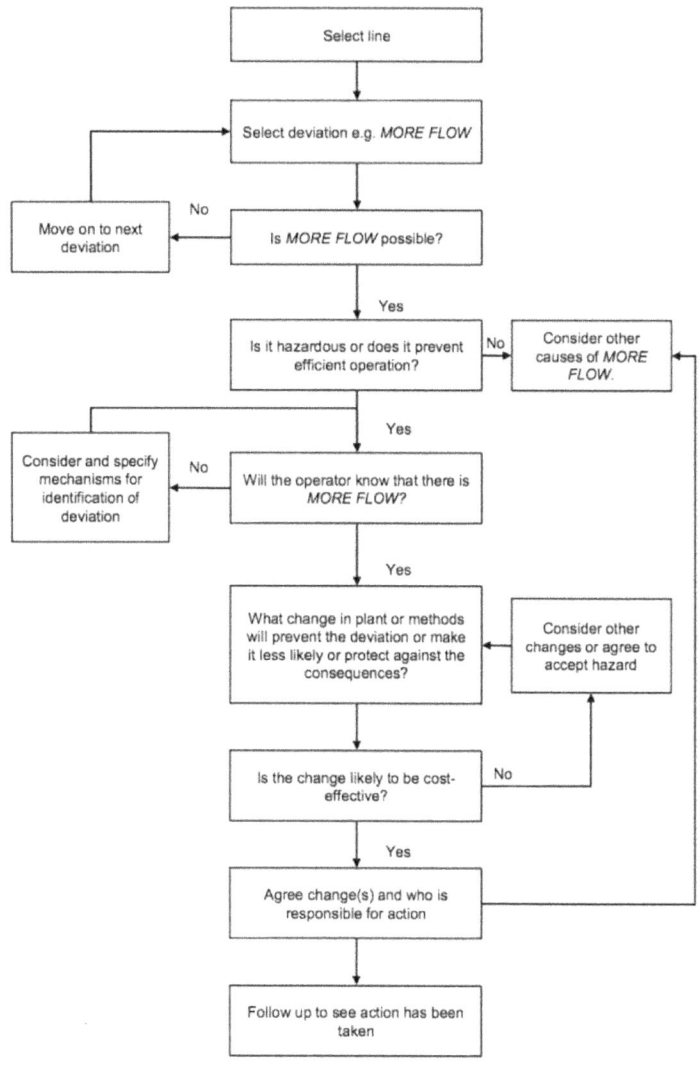

The following inputs are required by FSES in order to conduct the PHA Study: -

At a minimum the following information would be required in order to conduct the workshop:

- Existing PHA / HAZOP report
- P&ID's
- Cause and Effects Diagrams
- Facilities Design
- Operating Data and Procedures
- Maintenance Data and Procedures
- Interlock List
- Equipment Data Sheet

Persons required to attend the PHA Workshop:

As a minimum the following personnel would be required in order to conduct the workshop:

- Process Engineer
- Controls and Instrumentation Engineer
- Process Safety Engineer
- Maintenance representative
- Operations representative

Based on the project requirements, additional personnel may be required to attend the workshop, which will be highlighted within the ToR (Terms of reference).

The expected output of the PHA Study:

On award of the study FSES will issue a project ToR, which will highlight the assumptions that shall be made in the study, along with the workshop details,

6: Process Hazard Assessments (PHA):

methodology and data sources that will be utilised as well as any further information required from the client.

Upon acceptance of the ToR (Terms of reference), FSES will facilitate the PHA study through a workshop providing the facilitator and scribe. Once the workshop has been conducted FSES will prepare a PHA report describing the facility, the scope of work, a detailed methodology, the PHA worksheets, a summary of the PHA actions and any recommendations based on the discussions during the workshop. FSES highly recommend that following on from the PHA study, a SIL Determination analysis is conducted in order to determine the SIL requirements for the SIF's identified during the workshop.

A Process Hazard Analysis (PHA) research is designed to identify potential hazards in the way things are done at work. The intended outcome is a list of potential risks and suggestions for making things safer. It's similar to safety detective work in that we examine each step carefully to determine if anything could go wrong.

This list assists us in understanding what may cause accidents and how we may prevent them. So, the ultimate goal is to make our workplace safer for everyone by understanding the dangers and devising innovative strategies to prevent problems before they occur.

Chapter 7

ASTM, BS & ASME STANDARD

ASTM – American Society for Testing and Materials

ASTM International, known until 2001 as the American Society for Testing and Materials. It is an international standard organization that develops and publishes voluntary consensus technical standards for a wide range of materials, products, systems, and services. The organization's headquarters is in West Conshohocken, Pennsylvania.

The British Standards Institution (BSI) and American Society of Mechanical Engineers (ASME) are two well-known organizations that create and disseminate standards to guarantee the effectiveness, safety, and quality of goods and procedures in their respective areas. Below is a summary of each:

ASTM which was founded in 1898 as the American Section of the International Association for Testing and Materials, predates other standards organizations such as

BSI (1901), DIN (1917), ANSI (1918) and AFNOR (1926).

BS – British Standards

New Logo   Old Logo

British Standards (BS) is the standards produced by BSI Group incorporated under a Royal Charter as the National Standards Body (NSB) for the UK. It produces British Standards under the authority of the Charter.

Products and services certified by BSI having met the requirements of specific standards within designated schemes are awarded the Kite mark.

7: ASTM, BS & ASME Standard:

Overview:

- **Scope:**

  BSI is the National Standards Body of the United Kingdom, developing standards across various industries.

- **Headquarters:**

  BSI is based in the United Kingdom.

- **Key Areas:**

  BSI covers a broad spectrum of industries, including manufacturing, healthcare, information technology, and services.

- **Key Standards:**
  - BS EN series for European harmonized standards
  - BS ISO series for international standards adopted by BSI

Significance:

- BSI standards are influential not only in the UK but also in many countries around the world.
- BSI has been instrumental in contributing to international standards through its involvement in the development of ISO (International Organization for Standardization) standards.

ASME – AMERICAN SOCIETY OF MECHANICAL ENGINEERS

The American Society of Mechanical Engineers (ASME) promotes the art, science, and practice of multidisciplinary engineering and allied sciences around the globe through continuing education, training and professional development, codes and standards, research, conferences and publications, government relations, and other forms of outreach.

ASME is an engineering society, a standards organization, an R & D organization, a provider of training and education, and a non-profit organization. Founded as an engineering society focused on mechanical engineering in North America, ASME is today multidisciplinary and global. ASME has over 140,000 members in 158 countries worldwide.

- **Scope:** ASME is a professional association that sets internationally recognized standards for mechanical engineering.
- **Headquarters:** ASME is based in the United States.
- **Key Areas:** ASME standards cover a wide range of engineering disciplines, including pressure vessels, piping, nuclear power, elevators, and more.
- **Key Standards:**
    - ASME Boiler and Pressure Vessel Code (BPVC)
    - ASME B31 series for piping
    - ASME NQA-1 for nuclear quality assurance

7: ASTM, BS & ASME Standard:

Significance:

- ASME standards are widely adopted globally, especially in the United States and North America.
- Compliance with ASME standards is often a regulatory requirement for various industries.

Comparison of ASME and BSI:

- **Geographical Focus:**
    - ASME predominantly focuses on American and North American standards.
    - BSI serves as the UK's National Standards Body but has a global impact, especially in Europe.

- **Industry Coverage:**
    - ASME has a strong emphasis on mechanical engineering, pressure vessels, and nuclear technologies.
    - BSI covers a wide range of industries beyond mechanical engineering.

- **International Adoption:**
    - ASME standards are widely adopted globally, but their prominence is particularly strong in North America.
    - BSI standards have global recognition and are adopted in various regions.

- **Collaboration:**
    - Both ASME and BSI actively collaborate with international standardization bodies to contribute to the development of global standards.

In summary, ASME and BSI play crucial roles in setting standards that impact the safety, quality, and efficiency of products and processes.

While ASME is well-recognized in the American and North American context, BSI standards have a broader international influence, particularly in Europe.

Organizations operating in these regions often adhere to the relevant standards published by ASME or BSI to meet regulatory requirements and ensure the highest standards of safety and quality.

Chapter 8

EMERGENCY RESPONSE PLANNING

Emergency Response Planning means preparing for and handling unexpected incidents. It covers methods, resources, and activities to protect people, property, and the environment during an emergency.

This proactive approach provides a quick and coordinated response to limit the effects of unforeseen events, improving safety and successful crisis management.

An emergency plan specifies procedures for handling sudden or unexpected situations. The objective is to be prepared to: Prevent fatalities and injuries and to reduce damage to buildings, stock, and equipment.

4 Steps in the Emergency Planning Process

Industries can limit injuries and damages, and return more quickly to normal operations, if they plan ahead. The below table outlines the six steps recommended by the Federal Emergency Management Agency (FEMA) that

PROCESS SAFETY ENGINEER GUIDE

should be followed in setting up a comprehensive emergency management program.

1.	**Step 1 - Form a Planning Team** • Form the team. • Establish authority. • Issue a mission statement. • Establish a schedule and budget.
2.	**Step 2 - Understanding the situation** • Research and analyse information about potential hazards and threats. • If hazards and threats are seen as problems and operational plans are the solution, then hazard and threat identification and analysis are main steps in the planning process.

3	**Step 3 – Determine Goals and Objectives**
	
	• Based on the Hazard profile developed, the planning team thinks about how the hazard or threat would evolve and what defines a successful operation.
	• Beginning with a given intensity of the hazard or threat, the team plans for prevention and protection, through initial warning (if available), to its final impact (as identified through analysis) and generation of specific consequences (e.g., collapsed buildings, loss of critical services or infrastructure, death, injury, or displacement).
	• • Identify external resources.
	  • Conduct an insurance review.
	  • List potential emergencies.
	  • Estimate likelihood of each emergency.
	  • Assess the potential human impact.

8: Emergency Response Planning:

	• Assess the potential property impact. • Assess the potential business impact. • Assess internal and external resources. • Determine planning and resource priorities
4	**Step 4 - Develop the Plan** • Outline plan components. • Identify challenges and prioritize activities. • Write the plan. • Establish a training schedule. • Assign responsibility for training. • Coordinate plan with outside organizations. • Maintain contact with other corporate offices.
5	**Step 5 – Prepare, Review and Approve plan** • Review the plan and revise, as needed. • Seek final approval. • Distribute the plan. • Meet with outside groups. • Identify codes and regulations. • Identify critical products, services and operations The final draft prepared is circulated to organizations that have responsibilities for implementing the plan to obtain their comments.

6	**Step 6 - Implement and maintain the Plan**
	
	• Act on assessments recommendations.
	  • Integrate the plan into company operations Conduct training.
	  • Evaluate and modify the plan, as needed.
	• Evaluate the effectiveness of the plan

The actions taken in the initial minutes of an emergency are most critical. A prompt warning to employees to evacuate, move to shelter or lockdown can save lives. A call for help to public emergency services that provides full and accurate information will help to get the right responders and equipment.

An employee trained to administer first aid or perform CPR can be lifesaving. Action by employees with knowledge of building and process systems can help control a leak and minimize damage to the facility and the environment.

The first step when developing an emergency response plan is to conduct a Risk Assessment to identify potential emergency scenarios. An understanding of what can happen will help to determine resource requirements and to develop plans and procedures to prepare. The emergency plan should be consistent with the performance objectives.

Every facility should develop and implement an emergency plan for protecting employees, visitors, contractors and anyone else in the facility. This includes building evacuation ("fire drills"), sheltering from severe weather such as tornadoes, "shelter-in-place" from an exterior airborne hazard such as a chemical release and lockdown. Lockdown is protective action when faced with an act of violence.

In an emergency, the first priority should always be life safety. The second priority is the stabilization of the incident. There are many actions that can be taken to stabilize an incident and minimize potential damage. First aid and CPR by trained employees can save lives. Use of fire extinguishers by trained employees can extinguish a small fire. Containment of a small chemical spill and supervision of building utilities and systems can minimize

8: Emergency Response Planning:

damage to a building and help prevent environmental damage.

Some severe weather events can be forecast hours before they arrive, providing valuable time to protect a facility. A plan should be established and resources should be on hand, or quickly, available to prepare a facility. The plan should also include a process for damage assessment, salvage, protection of undamaged property and clean-up following an incident. These actions to minimize further damage and business disruption are examples of property conservation.

Guidance for the development of an emergency response plan can be found in this step. Build your emergency response plan using the Emergency Response Plan Worksheet.

Who should be involved in the planning process?

Individuals/Organizations	Inputs from the Planning Team
Senior Officers	Employee capable of identifying planning goals and essential tasks Should have Policy guidance and decision making capability
Emergency Manager	Having knowledge about all-hazards planning techniques Having knowledge about the interaction of the critical, operational, and strategic response

		levels
		Having knowledge about the prevention, protection, mitigation and response and recovery strategies Having knowledge about existing mitigation, emergency, continuity, and recovery plans
	Senior EMS designated officer	Knowledge about emergency medical treatment requirements

Specialized personnel and equipment resources

Knowledge about how EMS interacts with the Emergency Operations Centre and incident command |
| | Fire Chief | Knowledge about fire department procedures, on-scene safety requirements, hazardous materials response requirements, and search-and-rescue techniques

Knowledge about the fire-related risks

Specialized personnel and equipment resources |
| or | Police Chief designated | Knowledge about the prevention and protection |

8: Emergency Response Planning:

officer	strategies for the jurisdiction
	Knowledge about security strategies
	Specialized personnel and equipment resources
Hazardous Material Controller	Knowledge about hazardous materials that are produced, stored, or transported in or through the community
	Knowledge about hazards planning techniques
	Knowledge of current and proposed mitigation strategies
	Knowledge of existing mitigation plans

Expanded planning teams should include representatives from partners within the surrounding areas planning area, jurisdictions, and facilities or locations of concern and must include stakeholder organizations responsible for infrastructure, the economy, the environment, and quality of life.

Team Operation

Building a workable team takes time and effort and typically evolves through the following stages:

1. Forming: Individuals come together as a team. During this stage the team members may be unfamiliar with each other and uncertain of their roles on the team.

2. Storming: Brain storming is done on various aspects during which Team members become impatient, disillusioned, and may disagree.
3. Norming: Team members accept their roles and focus on the process.
4. Performing: Team members work together and make progress toward the goal.
5. Adjourning: Their task accomplished, team members may feel pride in their achievement.

Threat Analysis

Identify potential threats to your community. Threat analysis determines:

What can occur?

How often it is likely to occur.

The damage it is likely to cause.

How it is likely to affect the community.

How vulnerable the community is to the threat.

The steps in the threat analysis process are as under:

Identify threats.

Profile each threat.

Develop a community profile.

Determine vulnerability.

Create and apply scenarios.

8: Emergency Response Planning:

Threats can be classified into:

Natural: Natural threats tend to recur repeatedly in the same geographical region because they are related to weather patterns and/or physical characteristics of an area. Examples include: severe weather, fire, drought, typhoons, epidemics, etc. Some more examples of threats are as under:

Avalanche

Drought

Earthquake

Epidemic

Flood

Hurricane

Landslide

Tornado

Tsunami

Volcanic eruption

Wildfire

Winter storm

Technological: Technological threats originate from technological or industrial accidents, infrastructure failures, or certain human activities. Technological threats may include: cyber/database failures, fires, power failures, transportation accidents, bridge collapses, etc. Some other examples are:

- Airplane crash
- Dam failure
- Hazmat release

- Power failure
- Radiological release
- Train derailment

Human-Caused: Human-caused threats arise from deliberate, intentional human actions to threaten or harm the well-being of others. Human-caused threats may include: kidnappings, hostage situations, sabotage, civil disturbances, bombings, hijackings, terrorist acts, etc.

Profiling a Threat

Profiling a threat entails thoroughly examining its traits, motivations, and possible impact. Determine the threat's origin, techniques, and vulnerabilities. Consider its skills and aims in light of previous patterns. Assess the risk of harm to persons, assets, or information. This approach allows for the development of tailored countermeasures and preventive tactics to effectively minimize the danger, hence improving overall security preparedness.

Examine the threat's legitimacy and potential escalation scenarios as well. Consider external factors such as geopolitical pressures or socioeconomic situations that may have an impact on the threat's evolution. Continuous monitoring and information gathering contribute to a dynamic threat profile, allowing companies to adapt and build defences against emerging threats.

Update threat profiles on a regular basis based on evolving information and changes in the threat landscape. Collaborate with intelligence sources and use advanced analytics to gain a complete knowledge while assuring adaptable and effective security measures.

8: Emergency Response Planning:

Worksheet of Threat Profile:

Threat:
Potential Consequences: o Catastrophic (Mass fatalities/casualties, loss of governance and essential services, widespread damage) o **Severe** (Numerous fatalities/casualties, loss of essential services, and widespread damage) o Moderate (Limited number of fatalities/casualties and damage to property) o **Minor** (Little or no injuries and isolated damage)

Probability of Occurring: o High o Medium o Low	Past History: Has this type of incident occurred before? ❑ Yes ❑ No If yes, when? _

Areas Likely To Be most Affected :
Probable Duration:
Probable amount of warning time: • Minimal (or no) warning • 6 to 12 hours warning • 12 to 24 hours warning • More than 24 hours warning
Existing Population Warning Systems:

Some threats may pose such a limited risk to the community that additional analysis may not be necessary.

Determining Vulnerability

After compiling the threat profiles, it is important to quantify the community's risk so that the community can focus on the hazards that present the highest risk.

Risk is the predicted impact that a hazard would have on the people, services, and specific facilities in the community. For example, during heavy rains, a specific area might be at risk of flooding, leading to restricted access to critical facility in that area.

Quantifying risk involves:

- Identifying the elements of the community (populations, facilities, and equipment) that are potentially at risk from a specific threat.
- Developing response priorities.
- Assigning severity ratings.
- Compiling risk data into community risk profiles.

In surveying risk, it is helpful to develop response priorities. The following is a suggested hierarchy for setting priorities:

Priority 1: Life safety (including hazard areas, high-risk populations, and potential search and rescue situations). Keep in mind that response personnel cannot respond if their own facilities are affected.

Priority 2: Essential facilities.

Priority 3: Critical infrastructure (communication, utilities, and transportation systems).

8: Emergency Response Planning:

Prioritizing Risks

The community should assign each hazard a **severity rating**—or **risk index**— that will predict, to the degree possible, the damage that can be expected in that community as a result of that threat.

This rating quantifies the expected impact of a specific hazard on people, essential facilities, property, and response assets.

The following table on severity ratings may be used:

Severity	Characteristics
Catastrophic	- Multiple deaths. - Complete shutdown of critical facilities for 30 days or more. - More than 50 per cent of property severely damaged.
Critical	- Injuries and/or illnesses result in permanent disability. - Complete shutdown of critical facilities for at least 2 weeks. - More than 25 per cent of property is severely damaged.
Limited	- Injuries and/or illnesses do not result in permanent disability. - Complete shutdown of critical facilities for more than 1 week. - More than 10 per cent of property is

	severely damaged.
Negligible	• Injuries and/or illness treatable with first aid. Minor quality of life lost. • Shutdown of critical facilities and services for 24 hours or less. • Less than 10 per cent of property severely damaged.

Next, we must develop a risk index for each threat by assigning a value to each severity level (use the following values: 1 = catastrophic; 2 = critical; 3 = limited; 4 = negligible) for the following types of threat data.

- Magnitude.
- Frequency of occurrence.
- Speed of onset.
- Community impact (severity rating).
- Special characteristics.

Finally, average the severity level for each factor to determine the overall risk level for that threat.

Chapter 9

PROTECTIVE ACTIONS FOR LIFE SAFETY

In case of a hazard within a building such as a fire or chemical spill, occupants within the building should be evacuated or relocated safely. Other incidents such as a bomb threat or receipt of a suspicious package may also require evacuation. If a tornado warning is broadcast, everyone should be moved to the strongest part of the building and away from exterior glass. If a transportation accident on a nearby highway results in the release of a chemical cloud, the fire department may warn to "shelter-in-place." To protect employees from an act of violence, "lockdown" should be broadcast and everyone should hide or barricade themselves from perpetrator.

Protective actions for life safety include:

- Evacuation
- Sheltering
- Shelter-In-Place
- Lockdown

PROCESS SAFETY ENGINEER GUIDE

The emergency plan should include these protective actions.

Evacuation

Prompt evacuation of employees requires a warning system that can be heard throughout the building. Test your fire alarm system to determine if it can be heard by all employees. If there is no fire alarm system, use a PA (public address system) or other means to warn everyone to evacuate. Sound the evacuation signal during planned drills so employees are familiar with the sound.

Ensure that there are sufficient exits available at all times. Check to see that there are at least two exits from hazardous areas on every floor of every building. Building or fire codes may require more exits for larger buildings.

Walk around the building and verify that exits are marked with exit signs and there is sufficient lighting so people can safely travel to an exit. If you find anything that blocks an exit, have it removed.

Enter every stairwell, walk down the stairs, and open the exit door to the outside. Continue walking until you reach a safe place away from the building. Consider using this safe area as an assembly area for evacuees.

Appoint an evacuation team leader and assign employees to direct evacuation of the building. Assign at least one person to each floor to act as a "floor warden" to direct employees to the nearest safe exit. Assign a backup in case the floor warden is not available or if the size of the floor is very large. Ask employees if they would need any special assistance evacuating or moving to shelter. Assign a "buddy" or aide to assist persons with disabilities during an emergency. Contact the fire department to develop a plan to evacuate persons with disabilities.

9: Protective Actions for Life Safety:

Have a list of employees and maintain a visitor log at the front desk, reception area or main office area. Assign someone to take the lists to the assembly area when the building is evacuated. Use the lists to account for everyone and inform the fire department whether everyone has been accounted for. A fire, chemical spill or other hazard may block an exit, so make sure the evacuation team can direct employees to an alternate safe exit.

Sheltering

In case of a tornado warning, a distinct warning signal should be sounded and everyone should move to shelter in the strongest part of the building. Shelters may include basements or interior rooms with reinforced masonry construction. Evaluate potential shelters and conduct a drill to see whether shelter space can hold all employees. Since there may be little time to shelter when a tornado is approaching, early warning is important. If there is a severe thunderstorm, monitor news sources in case a tornado warning is broadcast. Tune in to weather warnings broadcast by local radio and television stations. Subscribe to free text and email warnings, which are available from multiple news and weather resources on the Internet.

Shelter-In-Place

You should develop a shelter-in-place plan. The plan should include a means to warn everyone to move away from windows and move to the core of the building. Warn anyone working outside to enter the building immediately. Move everyone to the second and higher floors in a multi-storey building. Avoid occupying the basement. Close exterior doors and windows and shut down the building's air handling system. Have everyone remain sheltered until public officials broadcast that it is safe to evacuate the building.

Lockdown

An act of violence in the workplace could occur without warning. If loud "pops" are heard and gunfire is suspected, every employee should know to hide and remain silent. They should seek refuge in a room, close and lock the door, and barricade the door if it can be done quickly. They should be trained to hide under a desk, in the corner of a room and away from the door or windows.

Incident Stabilization

Stabilizing an emergency may involve many different actions including: fire fighting, administering medical treatment, rescue, containing a spill of hazardous chemicals or handling a threat or act of violence.

When you dial 9-1-1 you expect professionals to respond to your facility. Depending upon the response time and capabilities of public emergency services and the hazards and resources within your facility, you may choose to do more to prepare for these incidents

Chapter 10

JHA/JSA (JOB HAZARD ANALYSIS)

A Job hazard analysis (**JHA**), also called a job safety analysis (JSA), is a technique to identify the dangers of specific tasks in order to reduce the risk of injury to workers. After knowing what the hazards are, we can reduce or eliminate them before anyone gets hurt.

We'll simplify Job Hazard Analysis (JHA) throughout this trip. It functions like a mega safety manual for all of your office chores. We'll get knowledge on how to identify potential threats, make our work safer, and maintain everyone's well-being. Prepare to comprehend and become an expert in the straightforward technique of safety!

It is important to know what jobs in your facility are hazardous.

Job hazard analyses are an effective way to identify where hazards exist in your facility and establish safe work practices

Before You Start a JHA

Before you begin the JHA for a specific job, do the following:

Get your employees involved

Safety works best when management **and** employees are both involved. That's true of the JHA process as well. Remember, it's their job, and they probably know it better than you do.

Review the history of injuries, illnesses, near misses and machine/tool damage

Go through written records of injuries, illnesses, near misses and incidents that have required machine/tool replacement or repair. Then, get feedback from your employees, asking if there are things that have occurred but are not in the records. (Make it clear you're trying to make work conditions safer, not punish anyone because something hasn't been reported.)

Ask the employees which hazards exist in their work area

Ask your employees if they're aware of hazards in their work area. Write them down — you can use this list later when you're performing the JHA.

Create a list that prioritizes the jobs for which you'll perform a JHA

It's great if you do a JHA for every job, but you should do JHAs for the jobs with the highest risks first. Take the information you've already gathered and prioritize the order in which you'll perform the JHAs.

With these steps down, you're now ready to complete the formal JHA process, described below.

How to Do a Job Hazard Analysis: 4 Essential Steps

Once you've completed the introductory steps above, it's time to begin the formal JHA process for a given job. Here's how to do a job hazard analysis:

10: JHA/JSA (Job Hazard Analysis):

Step 1: Begin the JHA for a specific job by breaking the job down into the steps or tasks performed while doing the job

Here are some ways to do this:

- Watch an employee performing the job
- Ask the employee what the various steps are — the employee may have some good insight here, but remember that the employee may leave out some steps because they're "automatic" to him or her
- Ask other employees who have performed the job to list or review the steps
- Film the employee while he or she performs the job — this will help you identify the steps

Write these steps down any way you want. It's common to create a JHA form that represents each task of a given job, plus a description of the task, the hazards and potential hazard controls.

Step 2: Identify and list the hazards associated with each task (do one task first, then another, etc.)

Danger detection: identify, evaluate, and take action. It's about identifying hazards, ensuring everyone's safety, and making wise decisions at work.

Consider every possible step that could go wrong. How could the worker be injured or be made ill? How could machines or equipment be damaged? Ask the following questions:

- What could go wrong?
- What could cause that thing/those things to go wrong?
- What other factors could contribute to that thing/those things going wrong?
- What would happen if that thing/those things did go wrong?
- How likely is it that that thing/those things will go wrong?

Step 3: Write a hazard description (also called a hazard scenario)

Write a description of each hazard in a consistent, orderly manner that will help ensure you will later put in steps to control the hazard and create the best possible controls.

A good hazard description should include the following items:

- **Environment**: Where does this hazard exist?
- **Exposure**: Who might be injured or made ill by this hazard?
- **Trigger**: What event might cause the hazard to lead to an injury or illness?
- **Contributing factors**: Are there other factors that might contribute to cause the hazard to lead to an injury or illness?
- **Outcome/consequence**: What would be the result if the hazard were to occur?
-

Step 4: Create a plan for controlling each hazard associated with each task

Once you've written the hazard description, it's time to brainstorm some hazard controls so the hazard never really does lead to an injury or illness. And remember what we said earlier - if you've identified a severe hazard and/or one with a great chance of causing illness or injury, address it immediately.

When you're considering a list of controls, think of the following (and in this order):

Elimination and/or substitution: If you can remove the hazard entirely, or put some form of substitute in place, do that. That's the best way to deal with a hazard. An example would be removing a sharp edge on the corner of a machine so nobody could get cut.

10: JHA/JSA (Job Hazard Analysis):

Engineering controls: Engineering controls involve redesigning the work area so that the hazard is eliminated or reduced. An example would be enclosing a noisy motor inside a soundproof box.

Administrative controls: Administrative controls involve modifying the way people work around a hazard to reduce the risk. An example might be limiting the number of hours someone works lifting heavy boxes from the end of a conveyor belt.

Personal protective equipment (PPE): PPE can be used to protect people who are working in the presence of hazards. An example would be giving a respirator to someone working near airborne crystalline silica. PPE should only be used as a last resort, once the other forms of controls listed above have been tried. PPE may be used in combination with the other forms of controls, also.

When should you control your hazards?

Once you've completed the JHA, start controlling them.

Should you review and revise your JHAs?

Yes, you should review your JHAs and, if necessary, revise them:

On a routine, periodic basis — maybe every year
When an injury or illness occurs at a specific job
When there's a close call or near miss
When the job changes

Four basic stages in conducting a JSA are:

Selecting the job to be analysed.

Breaking the job down into a sequence of steps.

Identifying potential hazards.

Determining preventive measures to overcome these hazards.

Is there a difference between a JHA and a JSA?

JSA stands for "Job Safety Analysis" while JHA stands for "Job Hazard Analysis." There is a minor difference between the two.

Both a JSA and JHA are a process to identify hazards within jobs and tasks and implement safety controls to reduce the risk of the potential hazards.

Both a job safety analysis and a job hazard analysis contain three specific components:

1. Looking at the steps that make up a job.

2. Identifying the hazards at each step.

3. Finding safety measures to avoid these risks.

The main difference between the two is a JHA also includes a risk assessment to evaluate potential hazards.

Since most safety professionals do include risk assessments in their JSAs. Most safety professionals utilize risk matrix calculations to assess and prioritize these hazards risks.

Ultimately, the end goals and results of a JSA and JHA are the same.

Does OSHA require a JSA?

OSHA does not specifically state that a JSA/JHA are required as part of this hazard assessment. Instead, it's important to know that performing and documenting a JHA or JSA will assist you in complying with the hazard assessment requirements.

10: JHA/JSA (Job Hazard Analysis):

What are the main features of a JSA Template?

A Job Safety Analysis (JSA) template is a formal document that outlines essential components for identifying, evaluating, and reducing hazards at work. Work steps, possible risks, suggested safety precautions, and emergency protocols are usually included.

By encouraging employees to recognize and rank the hazards connected to certain jobs, the template promotes a proactive safety culture. It might also include a signature section, indicating a dedication to security. A well-crafted JSA template acts as a useful manual, guaranteeing uniformity in hazard identification and encouraging on-going workplace safety improvement by offering a consistent framework for risk assessment and mitigation.

JSA Templates include the following:

- The job and associated tasks that make up the job;
- the identified hazards for each task / step;
- the hazard risk assessment;
- the hazard controls like PPE, Training etc.
- the mitigated risk assessment after the use of controls.

Organizations may also include on the template the date / time of the JSA history, as well as sign off and approvals from key individuals of the JSA.

Chapter 11

RISK ASSESSMENT

A risk assessment is a systematic examination of the workplace to:

1) identify significant hazards;

2) assess injury severity and likelihood; and

3) implement control measures to reduce workplace risks

Beyond complying with legislative requirements, the purpose of risk assessments is to improve the overall health and safety of the workers.

The key difference between a risk assessment and a JSA is scope. Risk assessments assess safety hazards across the entire workplace and are accompanied with a risk matrix to prioritize hazards and controls. Whereas a JSA focuses on job-specific risks and is typically performed for a single task, assessing each step of the job is equally important.

Here's a tabular comparison between Risk Assessment and Job Safety Analysis (JSA):

Aspect	Risk Assessment	Job Safety Analysis (JSA)
Focus and Scope	Evaluates broader organizational or project-wide risks.	Focuses on specific tasks or jobs, analyzing inherent risks associated with each step of the process.
Application	Applied across different organizational levels and various projects.	Primarily applied at the operational level, specifically for individual job tasks.
Time Frame	Often conducted at the beginning of a project or as part of strategic planning.	Typically performed before or during the execution of a specific job or task.
Scale of Analysis	Considers high-level risks, including financial, strategic, and operational aspects.	Focuses on micro-level risks associated with specific work activities, emphasizing immediate safety concerns.
Complexity	Involves a more complex analysis,	Generally simpler, breaking down tasks into

11: Risk Assessment:

Aspect	Risk Assessment	Job Safety Analysis (JSA)
	incorporating multiple variables and scenarios.	step-by-step analyses for easier implementation.
Perspective	Takes a holistic perspective, considering the overall impact on the organization.	Takes a task-specific perspective, concentrating on the safety aspects of individual job functions.
Documentation	Typically involves comprehensive documentation and formal reporting.	Involves documentation of specific job steps, hazards, and safety measures, often in a standardized form.
Frequency	Conducted periodically or as significant changes occur in the organization or project.	Conducted more frequently, often before each new job or task and whenever there are changes in procedures.
Integration with Safety Procedures	May influence the development of overarching safety policies and procedures.	Directly integrates with daily safety practices, providing actionable insights for immediate implementation.
Collaboration	Requires collaboration across different departments and management levels.	Involves collaboration between workers, supervisors, and safety personnel for specific job tasks.

This table provides a concise overview of the key differences between Risk Assessment and Job Safety Analysis.

How to Perform a Risk Assessment?

Risk assessments should be carried out by competent persons who are experienced in assessing hazard injury severity, likelihood and control measures. A new risk assessment should be carried out when there are new machines, substances and procedures which could lead to new hazards. They should be reviewed regularly and kept up to date.

Here are 5 steps to follow when performing a risk assessment at the workplace:

a) Identify hazards: Survey the workplace and look at what could reasonably be expected to cause harm. Check manufacturers or supplier instructions or data sheets for any obvious hazards. Review previous accident and near-miss reports.

b) Decide who might be harmed and how: Identify which group and demographic of workers might be harmed. Ask workers if they can think of anyone else who could be harmed by the hazard.

c) Evaluate the risks and decide on control measures: Look for existing controls in place. Follow the hierarchy of controls in prioritizing implementation of controls.

d) Record your findings and implement them: Use a risk assessment template to document your findings. Share your report and findings with key parties who can implement changes.

e).Review the assessment and update if necessary: Follow up with your assessments to check if controls have been put in place or if any new hazards have resulted

11: Risk Assessment:

Using a Risk Matrix

	Likelihood	Very Likely	Unlikely	Unlikely	Highly Unlikely
Consequences	Fatality	High	High	High	Medium
	Major Injury	High	High	Medium	Medium
	Minor Injuries	High	Medium	Medium	Low
	Negligible Injuries	Medium	Medium	Low	Low

A risk matrix is a graphical tool for assessing and ranking risks based on their likelihood and severity. It categorizes hazards into distinct levels using a grid, helping companies to focus on high-priority risks for effective mitigation and management, ensuring resources are spent where they are most required.

Risk matrix is often used during a risk assessment to measure the level of risk by considering the consequence/ severity and likelihood of injury to a worker after being exposed to a hazard.

The two measures can then help determine the overall risk rating of the hazard. Two important questions to ask when using a risk matrix should be:

Consequences:
How bad would the most severe injury be if exposed to the hazard?

Likelihood:

How likely is the person to be injured if exposed to the hazard?

How to Assess Consequences?

In assessing the consequences of a hazard, the first question should be asked "If a worker is exposed to this hazard, how bad would the most probable severe injury be?" For this consideration we are presuming that a hazard and injury is inevitable and we are only concerned with its severity.

It is common to group the injury severity and consequence into the following four categories:

- **Fatality-** leads to death

- **Major or serious injury-** serious damage to health which may be irreversible, requiring medical attention and on-going treatment

- **Minor injury-** reversible health damage which may require medical attention but limited on-going treatment). This is less likely to involve significant time off work.

- **Negligible injuries-** first aid only with little or no lost time.

How to Assess Likelihood?

In assessing the likelihood, the question should be asked "If the hazard occurs, what is the likelihood of worker getting injured." This should not be confused with how likely the hazard is to occur.

It is common to group the likelihood of a hazard causing worker injury into the following four categories:

Very likely – exposed to hazard continuously.

Likely – exposed to hazard occasionally.

11: Risk Assessment:

Unlikely – could happen but only rarely.

Highly unlikely– could happen, but probably never will.

How to Implement Control Measures?

After identifying and assigning a risk rating to a hazard, effective controls should be implemented to protect workers. Working through a hierarchy of controls can be an effective method of choosing the right control measure to reduce the risk.

Who should do risk assessments?

Risk assessments should always be carried out by a person who is experienced and competent to do so, competence can be expressed as a combination of Knowledge, Awareness, training, and experience. If necessary, consult a more experienced member of staff or external professional help to assist with the risk assessment template.

Remember competence does not mean you have to know everything about everything, competence also means knowing when you know enough or when you should call in further expert help.

When should risk assessment templates be done?

A separate risk assessment should be carried out for all tasks or processes undertaken by your organisation, they should be carried out before the task starts, or in the case of existing or long running tasks, as soon as is reasonably practicable.

Risk Assessments should also be reviewed on a regular basis; monthly, annually, bi-annually, depending on risk, or if something changes i.e., a new worker, a change of process or substance etc.

Non-Compliance

The penalties for failing to carry out risk assessments can be strict, The Process Health & Safety Officer can issue improvement or prohibition notices, this is likely to happen where he/she finds a situation with the potential to cause harm, for example an unguarded motor

What is a Method Statements?

A Safety Method Statement, sometimes called a "safe system of work" must be produced for all jobs or tasks that contain some measure of risk, contractors are more and more noticing that Method Statements are being requested by their clients, the request for a Method Statement can come at any time, Pre-Tender, Tender, Pre start of contract and sometimes after the contract has started, so it is best to be prepared

How to Complete a Risk Assessment?

Risk Assessment is often useful to fill in a template as per below 8 steps: -

1. Identify the hazards
2. Identify those at risk
3. Identify existing control measures
4. Evaluate the risk
5. Decide/Implement control measures
6. Record assessment
7. Monitor and review
8. Inform

Identify the hazards

A hazard is a situation or a condition with the potential for harm!

11: Risk Assessment:

Find out what the significant hazards associated with the task or processes are. There are several ways of identifying hazards; by observation, experience and talking to those who carry out the job you can also consult the following;

- Workforce
- Accident, ill health and near miss data
- Instruction Manuals
- Data sheets
- Hazard sheets
- Workplace inspections

Look for the hazards that you could reasonably expect to result in significant harm, for example;

Slipping and tripping hazards from poorly maintained floors, Fire hazards from flammable materials etc.

Identify those at risk

Think about individuals or groups of people who may be affected e.g.

- Office staff
- Maintenance personnel
- Members of the public
- Machine operators

Particular attention must be paid to disabled staff, lone workers, temporary staff and young inexperienced workers.

Identify Existing Control Procedures

Examine how you already control the risks; it is unlikely that your workers are getting injured on a daily basis, so you must have some controls in place already. To decide if those existing control procedures are adequate, and to evaluate the risk, complete a risk ranking which will determine the residual risk.

Evaluate the risk

A risk is defined as the likelihood that a hazard will cause harm.

Risk= Likelihood x Severity

Below is an example of a simple 1 to 5 risk ranking system.

1. Highly Unlikely
2. Unlikely
3. Possible
4. Probable
5. Certain

If the hazard does result in harm, how severe would the injury be?

1. Scratch (trivial)
2. Cut (minor injury)
3. Fracture (major injury-over 3 days injury)
4. Amputation (major injury)
5. Death ((Death)

To carry out a risk ranking simply multiply the likelihood by severity.

11: Risk Assessment:

The Likelihood-Severity table is as follows:

Likelihood	Very Likely	Unlikely	Unlikely	Highly Unlikely
Consequences — Fatality	High	High	High	Medium
Major Injury	High	High	Medium	Medium
Minor Injuries	High	Medium	Medium	Low
Negligible Injuries	Medium	Medium	Low	Low

| Likelihood | Severity | | | | |
	Trivial	Minor Injury	Over 3-day injury	Major Injury	Incapacity Or Death
Highly Unlikely	1	2	3	4	5
Unlikely	2	4	6	8	10
Possible	3	6	9	12	15
Probable	4	8	12	16	20
Certain	5	10	15	20	25

After the multiplication you will be left with a number from 1 to 25 which you can match against the following table to get the residual risk i.e., the risk that remains after the controls are in place.

The risk ranking will now give you your residual risk Low, Medium, or High.

If the risks are acceptable (Low Risk) then you may wish to skip the next part, if the risk are still Moderate/High (Medium/High Risk) then you must do something to bring the risk to a "tolerable" level, you can also prioritise your actions from 1 - 5.

Priority

1	Urgent Action - (Risk no 15 - 25)
2	High Priority - (Risk no 10 - 12)
3	Medium Priority - (Risk no 5 - 9)
4	Low Priority - (Risk no 2 - 4)
5	Very Low Priority - No Action required (Risk no 1)

Decide and implement new control measures

If the risk is not adequately controlled decide which new control procedures are required and ensure these procedures are implemented. The control measures are the actions performed to reduce either the probability of the accident happening or the severity of the outcome, and where possible both. When considering what measures to put in place it is important to consider both severity and likelihood, in order to minimise the overall risk.

11: Risk Assessment:

When deciding what new control measures will be required, it is helpful to work through the 'hierarchy' of controls. The hierarchy is as follows:

1. Elimination – get rid of the risk altogether
2. Substitution – exchange one risk for something less likely or severe
3. Physical Controls - separation/Isolation, eliminate contact with the hazard
4. Administrative controls - safe systems of work, rules in place to ensure safe use/contact with hazard
5. Information, instruction, training & supervision – warn people of hazard and tell/show/help them how to deal with it
6. Personal Protective Equipment – dress them appropriately to reduce severity of accident

Control measures should be practical and easy to understand (what to do and why they are doing it), applicable to the hazard, able to reduce the risk to acceptable levels, acceptable to the workforce and easy to operate.

After you have implemented the new control procedures, then re-rank the risks as above to determine the new residual risk, you should aim to get the risk to as low as is reasonably practicable until it is at a tolerable level.

Record the assessment

Keep copies of the assessments for your records and for inspection by the HSE should they ever be requested

Monitor and review

You must ensure that the control measures are achieving the desired level of control. You must review the assessment on a regular basis or if anything changes e.g. new staff, change in machinery or process.

Inform

You have a legal duty to relay the findings of the assessment to everyone who is affected by it. You must also provide information to the workforce on any new control measure implemented, any emergency procedures that have been developed and their duties as employees.

General

Get a few people to check your risk assessment template; other people may spot something that you have missed, also start a register of risk assessment so that you can find them quickly if needed.

What are the five steps to risk assessment?

Step 1: Identify hazards i.e., anything that may cause harm.

Employers have a duty to assess the health and safety risks faced by their workers. The Company must systematically check for possible physical, mental, chemical and biological hazards.

This is one common classification of hazards:

- Physical: e.g. lifting, awkward postures, slips and trips, noise, dust, machinery, computer equipment, etc.

- Mental: e.g. excess workload, long hours, working with high-need clients, bullying, etc. These are also called 'psychosocial' hazards, affecting mental health and occurring within working relationships.

11: Risk Assessment

- Chemical: e.g. asbestos, cleaning fluids, aerosols, etc.
- Biological: including tuberculosis, hepatitis and other infectious diseases faced by healthcare workers, home care staff and other healthcare professionals.

Step 2: Decide who may be harmed, and how.

Identifying who is at risk, whether own staff Contractors, visitors, clients and other members of the public on their premises.

Employers must review work routines in all the different locations and situations where their staff are employed. For example:

- Home cares supervisors must take due account of their client's personal safety in the home, and ensure safe working and lifting arrangements for their own home care staff.

Step 3: Assess the risks and take action.

This means employers must consider how likely it is that each hazard could cause harm. This will determine whether or not your employer should reduce the level of risk. Even after all precautions have been taken, some risk usually remains. Employers must decide for each remaining hazard whether the risk remains high, medium or low.

Step 4: Make a record of the findings.

Employers with five or more staff are required to record in writing the main findings of the risk assessment. This record should include details of any hazards noted in the risk assessment, and action taken to reduce or eliminate risk.

This record provides proof that the assessment was carried out, and is used as the basis for a later review of working practices. The risk assessment is a working document. You should be able to access it. It should not be locked away in a cupboard.

Step 5: Review the risk assessment.

A risk assessment must be kept under review in order to:

- ensure that agreed safe working practices continue to be applied (e.g. that management's safety instructions are respected by supervisors and line managers); and

- take account of any new working practices, new machinery or more demanding work targets.

Chapter 12

HIERARCHY OF CONTROLS

Hierarchy of Controls

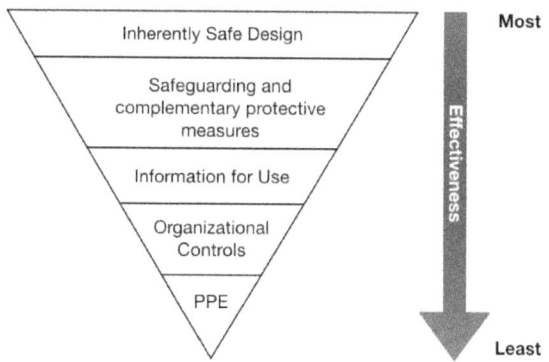

The Hierarchy of Controls is a strategic structure used in process management to prioritize and implement effective safety measures. It starts with removing or substituting dangers, then moves on to engineering controls, administrative controls, and personal protective equipment. This hierarchy assists organizations in systematically reducing hazards and promoting a safer workplace.

The Hierarchy of Controls are classified into:

- Engineering controls
- Management controls
- PPE (Personal protective equipment)

Engineering controls

Engineering controls in process management entail creating protections to reduce risks. Automated processes, enclosures, ventilation systems, and safety interlocks are some examples. These controls seek to physically alter the workplace environment or procedures in order to reduce the likelihood of risks and improve overall safety.

Redesigning the hazard out, ex. - Fabricate a mesh guard to protect against exposure to moving parts.

Replacing the unsafe item with a safe item Enclose the hazard

Substitute an unsafe item with different item. Ex. Substitute toxic chemical with a non-toxic chemical

12: Hierarchy of Controls:

Management controls

Manage work practices: Effective design and implementation of safe work procedures

Manage work schedules including job rotation, breaks, shift work, etc.

"Any system that depends on human reliability is inherently unreliable"

Personal protective equipment (PPE)

When engineering and/or administrative controls don't adequately eliminate or reduce the hazard(s) of a task, PPE may be needed in addition to those strategies.

PPE places a barrier between workers and the hazard. Remember, PPE does not eliminate or reduce the hazard itself, it merely sets up a barrier between you and the hazard.

Chapter 13

SAFE WORK METHOD STATEMENTS

SWMS or Safe Work Method Statement is a document that outlines high-risk construction work activities, hazards associated with the activity, and the control measures in place to manage these risks.

SWMS is required before performing any high-risk work to ensure safety and proper implementation of standard operating procedures.

Safe Work Method Statements (SWMS) are often a necessary tender requirement and where works defined as 'high risk work' are involved SWMS are mandatory.

There's just one problem — understanding how safe work method statements work can be confusing.

But since safe work method statements are a part of the legal framework a basic understanding of SWMS can help you and your team recognise what your obligations are.

- Is the SWMS primarily a 'statement' for a principal contractor to understand how a subcontractor will conduct their work practices safely on site?
- What is the legal requirement to have a Safe Work Method Statement (SWMS)?

The purpose of SWMS is to actually help supervisors and workers implement and monitor the control measures established at the workplace; however, it must also be provided to the principal contractor (if you are not already the principal contractor) before high-risk construction work starts.

The principal contractor must have a copy of sub-contractors SWMS before work starts for two practical reasons:

1. To determine beforehand that the sub-contractors documented Safe Work Practices are of an acceptable standard to allow for safe work on site and,
2. As a reference for those in charge of the workplace to ensure that the high risk work continues to be carried out in accordance with controls as stated in the SWMS.

The purpose of a SWMS is to enable all people involved with a full understanding of the risks involved with undertaking that work and to implement the risk controls outlined in the SWMS thereby increasing workplace safety.

What is the legal requirement to have a Safe Work Method Statement?

Legally, a Safe Work Method Statement is only required for work that is defined as 'high risk work'

In many jurisdictions, a Safe Work Method Statement (SWMS) is legally necessary to guarantee worker safety.

13: Safe Work Method Statements:

It must list the precise job duties, related risks, and safety precautions. Thorough risk evaluations, worker consultations, and adherence to pertinent health and safety standards are frequently necessary for compliance.

The document needs to be simple for employees to access, updated and reviewed frequently, and customized for the particular task at hand. Due to the potential for fines and threats to employees' general safety and wellbeing, noncompliance with these legal standards makes SWMS an essential component of risk management and regulatory compliance across a range of businesses.

Can I use a SWMS for work that is outside of the scope of the defined high-risk construction work?

Yes, the government has only specified what you must do to remain legally compliant. There is nothing that says you cannot develop and implement systems to the same level of care, or higher, for activities outside of the scope of the regulations.

For all other activities a SWMS is not required. However, a person conducting a business or undertaking must manage risks to health and safety by eliminating or minimising risks so far as is reasonably practicable."

In other words, you must still manage risks to health and safety by eliminating or minimising risks with all activities you undertake.

Here are 5 important things to include when creating a good SWMS report:

1. Break the job down – list the steps in a logical manner and take into consideration what is required to be achieved by the task

2. Identify the high-risk tasks – assess activity or task to which a person might be harmed or injured when working

3. Plan how to control – develop preventive and control measures to mitigate hazards i.e. what safety systems should we implement to make the job safer and prevent the injuries that may occur

4. Determine who is responsible – identify roles and responsibilities for actions and outcomes to make sure that controls are carried out and communicated properly.

5. Educate workers – ensure the SWMS is fully understood by all workers prior to commencing the task.

A SWMS should be made available for inspection at any given time. It must also be reviewed each year and amended if necessary.

Task Analysis/Safe Work Method Statement

For certain types of work, you'll need to show you've planned ahead. Part of our Site-Specific Safety Plan (SSSP), a Task Analysis/Safe Work Method Statement is an excellent planning tool that ensures all risks and controls are identified - and helps improve productivity by ensuring the right people, plant and processes are ready when you need them. A TA should describe how you plan to do the job safely and proves that you are managing risk.

First things first – do you need to complete a TA?

TA/SWMS are required for:

- Particular risks specified by legislation
- Any new or complex task
- When required by contract

13: Safe Work Method Statements:

- Permit to Work Systems
- Work requiring Certificate of Competence
- If your risk assessment results in CRITICAL or HIGH level or risk
- Notifiable construction work

If you do need to complete a TA, simply follow the steps below:

1. The job details
2. The people
3. Step-by-step
4. Identify hazards and assess risk
5. Control risks
6. Reassess risks

Step 1: The job details

Complete top left fields with your company and site name, and date of task. Complete (your) subcontractor details, your name and date TA was written. Complete all details, including personal protective equipment required e.g., safety boots and administrative controls e.g., traffic management plan, pre-start meetings, hazard boards.

Step 2: The people

Record names of all workers involved in task to show they understand requirements of this TA. Workers should sign beside their name to show they have been involved in the TA/SWMS and understand it.

Step 3: Step-by-step

Break down the task into the major steps from first to last and record them on the TA. There will normally be at least three to four steps but may be more if it is a complex task e.g., step 1 = prestart inspection, step 2 = set up site, step 3 = complete task, step 4 = clean-up site etc.

Step 4: Identify hazards and assess risk

Record all the hazards you might encounter during each step e.g., 1a = other traffic/plant entering site. Complete an initial risk assessment using the risk matrix and record the risk level e.g., Critical, high, moderate, low, very low etc.

Step 5: Control risks

Record the controls used for each hazard and the level of control using the hierarchy of controls.

Step 6: Reassess risk

Go back to the risk matrix and work out the new/residual level of risk that you could expect with the above controls in place e.g., critical, high, moderate, low, very low. If you still get a "critical" level or risk, review your controls. Record the new level of risk for each hazard under each step.

Repeat the above steps until you have recorded the hazards, controls risk levels for each step of the task.

Chapter 14

LAYER OF PROTECTION ANALYSIS (LOPA)

LOPA is a semi-quantitative tool to establish adequacy of the protection layers against a hazard.

This methodology uses Independent Protection Layers (IPLs) along with other modifiers to provide an order-of-magnitude estimation of the likelihood of a given consequence.

Risk acceptance and reduction, if necessary, can then be computed using corporate risk standards. If the analysis determines that additional risk reduction is required, a Safety Instrumented Function (SIF) may be identified among other safeguard considerations. LOPA provides the tool to establish the necessary Safety Integrity Level (SIL) of the SIF, which is calculated through the SIL Assignment process.

Enhancing Safety in Complex Processes

Layer of Protection Analysis (LOPA) is an effective methodology that systematically aligns layers of defense to mitigate potential hazards in the complex world of process safety.

LOPA provides an organized method for evaluating the effectiveness of current safety measures and pinpointing the extra layers that are required to improve security.

Key Principles:

LOPA is based on the core idea that in order to prevent or lessen the effects of a dangerous occurrence, several levels of protection, or safeguards, are necessary.

These layers include administrative controls, human factors, safety instrumented systems (SIS), and intrinsic process safety features.

Risk Assessment:

The procedure starts with a thorough risk assessment that evaluates the likelihood and severity of probable incidents. This crucial study lays the framework for identifying the appropriate risk reduction targets and protective layers.

Assigning Risk Reduction Credits:

Each safeguard is given a risk reduction credit based on how effective it is at mitigating the specified dangers. These credits are quantifiable in LOPA, allowing for a systematic evaluation of cumulative risk reduction with predetermined targets.

14: Layer of Protection Analysis (LOPA):

Determining Risk Tolerance:

LOPA involves identifying an acceptable level of risk, which is frequently governed by industry norms or legal regulations. This stage ensures that the calculated risk reduction fits with stated safety goals, establishing a decision-making framework for the selection and execution of safeguards.

Benefits of LOPA:

1. **Efficient Resource Allocation:** LOPA allows businesses to optimally spend resources by focusing on putting safeguards where they are most needed, maximizing the balance between safety and operational factors.
2. **Clear Risk Communication:** LOPA enables straightforward communication of risk scenarios, allowing stakeholders to comprehend the potential repercussions of hazardous events as well as the effectiveness of existing measures.
3. **LOPA enables straightforward communication** of risk scenarios, allowing stakeholders to comprehend the potential repercussions of hazardous events as well as the effectiveness of existing measures.
4. **Decision Support Tool:** It is a decision-making tool that aids in the selection of appropriate risk-reduction strategies and the justification of investments in safety systems.
5. **Regulatory Compliance:** LOPA assists firms in meeting regulatory requirements by identifying and managing risks in a methodical manner, displaying a commitment to safety and due diligence.
6. **Continuous Improvement:** Organizations cultivate a continuous improvement culture by periodically evaluating and revising LOPA assessments, adjusting to modifications in procedures, technology, and industry norms.

Challenges and Considerations:

LOPA has advantages, but it also has drawbacks, like the requirement for expert opinion and the arbitrary allocation of risk reduction credits. The approach necessitates a knowledgeable team with experience in process safety as well as in-depth knowledge of the particular sector and processes being evaluated.

To sum up, Layer of Protection Analysis is a versatile and strong instrument for process safety. Through a methodical analysis of risks, a quantification of protections and compliance with regulatory standards, LOPA enables enterprises to make well-informed decisions that improve safety, preserve resources, and ensure employee wellbeing. It acts as a lighthouse, pointing industries in the direction of a future in which risks are reduced by means of strategic defensive layers.

You can go ahead to conduct:

LOPA for every major vessel in all process units of a major petroleum refinery.

LOPA for high severity consequences in all process units of a major petroleum refinery.

LOPA and SIL Assignment tasks for systems during the Front-End Engineering Design (FEED) phase of a major chemical manufacturing facility and refinery units.

Chapter 15

FEED

The crucial first stage of project development is called front-end engineering design, or FEED, and it involves turning abstract concepts into intricate technical plans. This phase addresses the technical, financial, and environmental aspects of the project and sets the foundation for its successful completion.

FEED guarantees a thorough comprehension prior to moving forward with the phases of detailed design and construction.

Engineers evaluate a project's viability, pinpoint any obstacles, and specify design specifications at the FEED stage. It entails risk assessments, initial cost estimates, and building a strong foundation for later project phases. In order to reduce uncertainty, maximize design concepts, and create a strong foundation for a project's effective execution, FEED is essential.

Front-end engineering and design are required to produce quality process and engineering documentation of sufficient depth, defining the project requirements for detailed engineering, procurement, fabrication and construction of facilities and supporting a ±10 per cent project cost estimate. This distinct project phase is typically used to:

- Conduct hazardous operations reviews
- Develop the engineering design packages that can be used to bid a lump sum EPC scope and / or provide the foundation for the detailed engineering phase
- Evaluate options that will improve the return on assets (ROA)
- Prepare cost estimates for scope definition and for project funding
- Support internal funding requirements

FEED is more than a cost estimate

A FEED analysis is much more than a cost estimate. Specifically, it provides the groundwork and technical detail from which a project is built. The basic engineering decisions made during the FEED have a significant impact on every project phase that follows.

Some organizations minimize their up-front engineering investments, either because they don't see the value in FEED or because of time or cost restraints. Skipping or minimizing this step, though, can limit their ability to accurately define project scopes, result in decisions that are based on assumptions and lead to price estimates that are made in haste – all of which can actually increase costs in the long run.

15: FEED:

In short, those involved in capital building projects should remember that FEED is project design that leads to a project cost – not the other way around.

Partner expertise is the key:

Chemical producers have long relied on engineering, procurement and construction (EPC) firms to conduct FEEDs. Those EPCs – much like their end-user customers – are facing a skills gap as their engineers retire. At the same time, industrial automation is increasingly connected and complex. This is why chemical producers should choose FEED partners with relevant automation expertise.

A control system may represent a fraction of the total project cost, but it is integral to the overall operation and respective goals. For instance, a chemical site will not receive the right recoveries or achieve the payback expected on paper if the system is not controlling operations at an optimal rate.

Because of this, automation solution providers are playing a growing role in carrying out FEED studies in tandem with EPC firms. In addition, using a single solution provider instead of multiple vendors can be more cost effective and efficient while reducing risk and engineering costs.

Some questions to ask when evaluating a vendor include:

- Do they speak the language and technical jargon that workers speak?
- Do they understand the applications and environments?

- Are they familiar with the equipment, safety regulations and technology trends?

Considering if the automation partner can help a chemical producer understand the risks and rewards of implementing such a technology in its facility is important, especially as operations become more connected

Early stakeholder involvement is important

Good FEED requires support and involvement from a cross-functional team within an organization. This should include the engineering, finance, operations, regulatory and facility-management teams.

Early stakeholder buy-in, including securing stakeholders' commitments to the process and confirming their requirements, aids in ensuring up front project agreement.

After a stakeholder team is established, a lead FEED contact should be designated to bring focus to the process. This person can help maintain stakeholders' involved throughout, such as with collecting their input and ultimately confirming that the FEED plan meets their specific requirements

Deliverables will vary

While FEEDs are equally important for both Greenfield (new build) and brownfield (rebuild, upgrade or extension) projects, the deliverables will differ slightly between them.

Most of the deliverables will be generated from scratch and that the FEED can directly lead into engineering the solution.

Brownfield projects, on the other hand, require more effort at the onset to document what is currently installed and to confirm the information that will be used

15: FEED:

as the basis from which the project can start. As a result, obsolescence, age of spare parts and probability of irreparable failure reports are needed, as are site surveys.

Brownfield projects also require clear commissioning and qualification strategies for any new processes being introduced to the existing facility, particularly for highly regulated industries.

Looking ahead today can save costs tomorrow

A chemical producer may be tempted to implement the most cost-effective production system that solely satisfies its immediate needs. However, operations evolve. Chemical producers should consider the time and financial costs involved in reinvesting to replace their system in a few years.

The more practical approach is to invest the time and resources up front, during the basic-engineering phase, to examine how a system will need to evolve as a facility evolves. The FEED analysis can help identify a facility's requirements today, but it also should include research into what will happen in the years ahead.

Building a good outcome

A chemical production site's value is defined and built during the FEED stage, long before production begins. A good FEED can help determine a project's usability; performance; cost effectiveness; and long-term operability, safety and environmental compatibilities.

Chapter 16

WHAT IS AN EER DIAGRAM?

The Energy Efficiency Ratio (EER) is a measure of an air conditioning system's or appliance's efficiency, notably in the context of cooling. Under specified working conditions, it is defined as the ratio of cooling capacity (in British Thermal Units or BTUs) to power input (in watts).

Understanding EER:

The EER is calculated using the following formula:

$$EER = \frac{\text{Cooling Capacity (BTUs)}}{\text{Power Input (Watts)}}$$

In simpler terms, it quantifies how effectively an air conditioning unit can cool a space relative to the amount of electrical energy it consumes.

For example, the **EER** rating for an **air conditioner** is calculated by dividing the BTU (British Thermal Units) rating by the wattage. For example, a 12,000-BTU **air conditioner** that uses 1,200 watts has an **EER** rating of 10 (12,000/1,200 = 10).

The air conditioner **EER** is its British thermal units (BTU) **rating** over its wattage. The **higher** the **rating** is, the more efficient the air conditioning unit is.

Significance:

1. **Efficiency Benchmark:** EER is used by consumers and industry professionals to evaluate the energy efficiency of cooling units. Higher EER levels imply greater efficiency.
2. **Cost Savings:** Appliances with higher EER ratings are generally more energy-efficient, resulting in cheaper long-term electricity expenses for consumers.
3. **Environmental Impact:** Energy-efficient appliances help to reduce energy consumption, which helps to reduce greenhouse gas emissions and the environmental effect of energy production.

Interpreting EER Ratings:

- **Standard EER Ratings:** The energy efficiency ratio (EER) of air conditioning units is frequently marked on the label. EER ratings for standard residential units typically range from 8 to 12.
- **Higher is better:** Better energy efficiency is indicated by a higher EER. An air conditioner with an EER of 12, for instance, uses less energy than one with an EER of 10.
- **Regulatory Standards:** Certain regions have legislative standards that establish minimum EER

16: What is an EER diagram?:

criteria for appliances sold on the market, thereby encouraging energy efficiency and environmental sustainability.

Factors Influencing EER:

1. **Design and Technology:** An air conditioner's design and technological elements have a big influence on its efficiency ratio (EER). Higher EER values are a result of advanced technology like energy-efficient heat exchangers and variable-speed compressors.
2. **Size and Capacity:** An air conditioner's efficiency is influenced by its size and cooling capacity. For best results, the unit's size must correspond with the area it is cooling.
3. **Maintenance:** Over time, maintaining the appliance's efficiency requires routine maintenance, which includes checking the levels of refrigerant and cleaning filters.
4. **Climate Conditions:** EER values are frequently calculated using particular humidity and temperature parameters. The appliance's actual performance could differ depending on the environment in which it is used.

Considerations for Consumers:

1. **Cost vs. Efficiency:** Although appliances with higher EER ratings are generally more efficient, consumers should consider both the appliance's initial cost and the potential for long-term energy savings.
2. **Climate Considerations:** The environment in which the appliance will be utilized should be taken into account. Choosing an energy-efficient unit can save a lot of money, especially in hot areas where air conditioning is often necessary.
3. **Energy Labels:** Many nations mandate energy labeling on appliances, including air conditioners, to

educate consumers about the energy efficiency of the device, including its EER rating.

To sum up, one essential statistic for assessing the energy efficiency of air conditioning systems is the Energy Efficiency Ratio, or EER. Customers are better able to make decisions that support environmental sustainability, lower utility costs, and meet energy efficiency targets when they are aware of and take into account EER ratings.

How to Draw ER Diagrams

1. Identify all the entities in the system. An entity should appear only once in a particular diagram.
2. Identify relationships between entities. Connect them using a line and add a diamond in the middle describing the relationship.
3. Add attributes for entities.

ER diagram is a visual representation of data based on **ER model**, and it describes how entities are related to each other **in the** database.

EER diagram is a visual representation of data, based on **EER model** that is an extension of the original entity-relationship (**ER**) **model**.

The process of creating an Enhanced Entity-Relationship (EER) diagram entails putting a database's entities, relationships, characteristics, and constraints on paper.

The following is a step-by-step tutorial on drawing an EER diagram (Please note that for ease of understanding a very simple example of "Student", "Course" and "Instructor" is used. Later in this section 2 real life Industry examples are elaborated):

16: What is an EER diagram?:

1. Identify Entities:

Start by figuring out which entities are in your database. Entities are things or ideas that have information kept about them. An instance of an entity in a university database would be "Student," "Course," and "Instructor."

2. Define Attributes:

Identify and describe the attributes of each entity. The qualities or features of the things are called attributes. A "Student" entity may have the following attributes: "Name," "DateOfBirth," and "StudentID."

3. Identify Relationships:

Establish the connections between the various entities. Relationships specify the connections between entities. A "Course" object and a "Student" entity, for example, may be related, showing that a student enrolls in courses.

4. Define Relationship Cardinality:

Specify the cardinality of each relationship. Cardinality indicates the number of instances of one entity that can be connected with another. Common notations include "1:1" (one-to-one), "1:N" (one-to-many), and "N:M" (many-to-many).

5. Determine Attribute Types:

Identify attribute types such as main keys, foreign keys, and composite attributes. Each record in an entity is uniquely identified by a primary key, whereas foreign keys connect entities in relationships.

6. Use Specialization and Generalization:

Use specialization and generalization to depict hierarchies within entities. For example, the general entity "Person" can be subdivided into "Student" and "Instructor."

7. Incorporate Weak Entities:

Include weak entities (entities that cannot be uniquely identified by their attributes alone) in your database diagram.

8. Draw the EER Diagram:

Draw the entities, characteristics, relationships, and cardinality using standard symbols. Rectangles for entities, ovals for characteristics, lines joining entities to depict relationships, and diamonds to denote the cardinality of relationships are all common symbols.

9. Annotate Relationships and Attributes:

Label relationships and properties to provide context and clarity. Include information on data kinds, constraints, and any other facts that will help the reader comprehend.

10. Validate and Refine:

Verify that your EER diagram appropriately reflects the requirements of your database. Consider possible scenarios and make sure the diagram supports them. To improve clarity and completeness, edit the diagram as needed.

16: What is an EER diagram?:

Tools for Drawing EER Diagrams:

To generate EER diagrams, use specialized software tools such as lucid chart, draw.io, or Microsoft Visio. These tools frequently provide predefined shapes and connectors, which makes drawing and modifying your diagram easier.

Tips:

- Keep the diagram clear and concise. Use consistent naming conventions for entities and attributes.
- Review and refine the diagram collaboratively with stakeholders to ensure accuracy and completeness.
- Consider the normalization process to reduce data redundancy and improve database efficiency.

Following these instructions will allow you to produce a detailed and accurate EER diagram that will serve as a visual representation of your database structure.

Creating full Process Safety Enhanced Entity-Relationship (EER) diagrams often entails illustrating complicated relationships within the context of a process or system. However, for illustration purposes, consider two simplified Process Safety scenarios:

EXAMPLES FROM INDUSTRY

Example 1:

Chemical Manufacturing Process Safety Database

Entities:

- Hazardous Substance
- Process Unit

- Safety Instrumented System (SIS)
- Incident

Attributes:

- Hazardous Substance: SubstanceID (Primary Key), Name, Chemical Formula
- Process Unit: UnitID (Primary Key), Name, Capacity
- SIS: SIS_ID (Primary Key), Type, Safety Integrity Level (SIL)
- Incident: IncidentID (Primary Key), Date, Description

Relationships:

- Handles: Connects Process Unit to Hazardous Substance (Many-to-Many)
- Implements: Connects Process Unit to SIS (One-to-Many)
- Reports: Connects Process Unit to Incident (One-to-Many)

This EER diagram represents the relationships between hazardous substances, process units, safety instrumented systems, and incidents in a chemical manufacturing context.

Example 2: Oil Refinery Safety Management System

Entities:

- Equipment
- Process Area
- Emergency Shutdown System (ESD)
- Safety Audit

16: What is an EER diagram?:

Attributes:

- Equipment: EquipmentID (Primary Key), Type, Location
- Process Area: AreaID (Primary Key), Name, Description
- ESD: ESD_ID (Primary Key), Type, SIL
- Safety Audit: AuditID (Primary Key), Date, Auditor

Relationships:

- Located In: Connects Equipment to Process Area (Many-to-One)
- Monitors: Connects ESD to Equipment (One-to-Many)
- Conducts: Connects Process Area to Safety Audit (One-to-Many)

This EER diagram represents relationships between equipment, process areas, emergency shutdown systems, and safety audits within the context of an oil refinery.

Please note that these examples are simplified and intended for illustrative purposes. A complete EER diagram for Process Safety would involve more detailed entities, attributes, and relationships specific to the processes and systems within a given industry or facility.

Chapter 17

SAFETY INTEGRITY LEVEL (SIL)

SIL Determination

While the Hazard and Operability (HAZOP) study identifies and risk ranks hazards, Safety Integrity Level (SIL) Determination focuses on the adequacy of safeguards to mitigate hazards.

Within the framework of a HAZOP, analysts are restricted to the limits of the governing risk matrix (i.e. specific range limits on frequencies of occurrence). In contrast, SIL analysis enables analysts to refine the estimates of frequencies of occurrence to obtain more realistic estimates of risk.

SIL Determination is the process of determining the amount of risk mitigation required to reduce the risk put forth by a process to a tolerable level. SIL Determination is the first step in the development, design, commission and operation of a Safety Instrumented System (SIS).

SIL Determination involves the determination of the safety integrity level (SIL) for each Safety Instrumented Function (SIF) in a Safety Instrumented System and is dependent on the following factors:

Corporate standards for the tolerable risk after applying all the layers of protection

The overall risk that can occur from the unprotected hazards.

The risk reduction provided by all of the non-SIS protection layers

The ideal time for SIL Determination to be done is during the front-end engineering design (FEED) and project definition stages and, typically, as a supplement to the HAZOP. An effective risk reduction and protection layer design draws on the same information and personnel involved in the initial hazard study (HAZOP).

Nevertheless, SIL Determination may also be used effectively during the plant's life to determine if improvements are needed and to provide guidance as to the form of the improvements.

The initial protection layer design may need to be reviewed and changed soon after the HAZOP has been completed. If the protection layer design and SIL Determination is delayed for some considerable time after the HAZOP, there is the risk that the SIS will have the feel of a "bolted on" solution and the capability of being able to integrate SIS and non-SIS protection layers may be compromised.

Careful planning and management is needed in the lifecycle of a SIS. However, while the SIS Safety Life Cycle model provides guidance on the steps necessary for a successful SIS project.

The determination of SIL is driven by company decision criteria such as Tolerable Frequencies, ALARP and

17: Safety Integrity Level (SIL) :

corporate risk tolerance philosophy. Standards such as IEC 61511 and OSHA's Process Safety Management legislation require the process industry to use good engineering practice in the design and operation of their facilities. This means that the determination of safety integrity levels must be competently performed and properly documented.

SIL Concept Validation

SIL Concept Validation can be done immediately following SIL Determination. It starts with situations where unmitigated risk exists and involves the design of one or more conceptual solutions that would mitigate the residual risk. The specified safeguards will mitigate all risk according to corporate risk targets.

It is also called "Concept Validation" because when the actual plant is built, the components (i.e. logic solvers, final elements, sensors) incorporated into the SIF need to be validated so the user confirms that the plant is properly protected

Safety Requirement Specification (SRS)

The SRS develops specifications for safety functions, including a Safety Instrumented System (SIS) design, and includes tags, functionality, performance and physical requirements. It is developed following SIL Determination and SIL Concept Validation to ensure that the critical control and safety system is designed to meet the necessary technical requirements. The SRS defines the functional and integrity requirements of the critical control system and serves as the basis to begin detailed engineering and programming of the SIS hardware and software.

The SRS is developed during the execution of a project involving a SIS and provides a key measure by which the SIS design is compared to and judged throughout its life cycle. It acts as a living document for the life of the facility.

SilCoreTM Software

SilCoreTM is a field proven, IEC compliant SIL Life Cycle tool that gives high integrity and critical control systems designers, engineers, operators and maintainers the information and power to conduct SIL Determination, SIL Validation and SIL Optimization exercises. It is used in the execution of the all-ACM SIL studies.

Typical Risk Reduction methods in Process plants

The process of Risk reduction in Plants is depicted in the following diagram for ease of understanding.

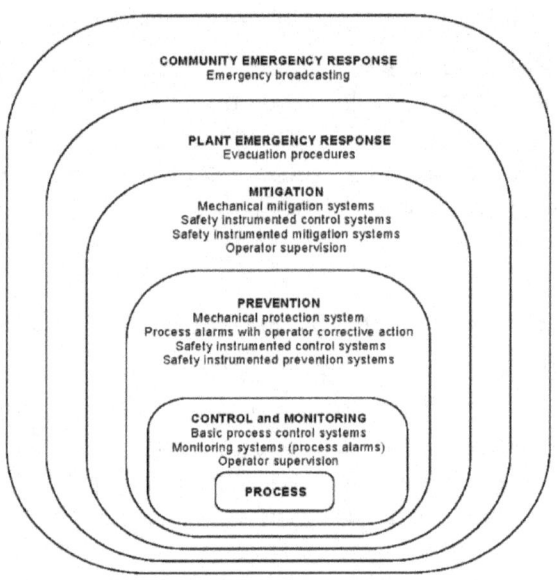

Typical risk reduction methods found in process plants (for example, protection layer model)

SIL DETERMINATION TECHNIQUES

The IEC 61511 standard refers to six SIL Determination techniques

ALARP and Tolerable Risk Concepts

17: Safety Integrity Level (SIL) :

Semi-Quantitative Method – Event Tree Analysis

Safety Layer Matrix Method

Calibrated Risk Graph

Risk Graph

Layer of Protection Analysis LOPA

ALARP and Tolerable Risk Concept

The ALARP (As Low as Reasonably Practicable) principle sets the stage for doing SIL Determination and is not used to actually determine SIL levels. The ALARP principle helps to define the tolerable risk target for a facility in terms of the social, political and economic factors and predefined consequences relevant to the company.

ALARP is a fundamental requirement for the management of industrial risks. The risk interpretation from ALARP is developed into several other forms. These include Risk Matrix (for HAZOP) as well as SIL Risk Matrix (Safety Layer Matrix), Calibrated Risk Graph and Layer of Protection Analysis (LOPA) definitions for SIL Determination.

Semi-Quantitative Method – Fault Tree and Event Tree Analysis

Fault trees and event trees are quantitative methods used to determine the frequencies of hazardous events. These frequencies may then be compared to a pre-defined Tolerable Frequency (TF).

The event tree displays the demand rate of all the initiating events resulting in the same consequence. Calling fault tree and event tree methods "semi-quantitative" could be perceived as a misnomer. Both methods are quantitative and the combination of the two is a powerful and rigorous method for determining SIL. Fault tree and event tree analysis often requires the use of specialized, quantitative risk assessment software.

Advantages

- Objective
- Graphically easy to understand
- Very powerful
- Mathematically rigorous

Safety Layer Matrix

Disadvantages

Requires skill in probabilistic methods to apply properly

The Safety Layer Matrix method identifies the risk reduction or SIL required. It is often referred to as the "SIL risk matrix" approach.

The safeguards are evaluated using Independent Protection Layer (IPL) rules defined by the company.

The Safety Layer Matrix is applied to the most frequent initiating event resulting in the same consequence. This results in a qualitative analysis on which initiating event has the highest frequency of occurrence. It is occasionally used in the process industry and can be used for integrated HAZOP/SIL determination sessions.

Advantages

Non numerical
HAZOP/ SIL matrix at same session
Easy to understand

Calibrated Risk Graph

The Calibrated Risk Graph is set by the company to meet the intent of ALARP and the related tolerable

17: Safety Integrity Level (SIL) :

frequencies. Calibration of the risk graph is the process of assigning numerical values to the risk graph variables. This forms the basis of assessment of the risk and determination of the required integrity of the SIF under consideration.

The Calibrated Risk Graph is applied to the most frequent initiating event resulting in the same consequence. This may result in some qualitative analysis to determine which initiating event has the highest frequency of occurrence for a specific consequence. The Calibrated Risk Graph approach normally involves three graphs - Safety, Environmental Impact, and Asset/Production Loss.

This SIL Determination method often requires more time than others because of the multiple evaluations of consequence categories.

Advantages

- Non numerical approach
- Easy to apply
- Commonly used when SIL first applied

Disadvantages

Qualitative, does not yield well-defined numerical estimates of risk

Time consuming on ranking the risk

Chapter 18

HSE AUDIT

HSE Auditing is a Systematic Examination involving Analysis, Tests, and Confirmations of Operational Procedures and Practices to verify. Whether they conform and comply with Legal Requirements and Internal Policies and evaluate whether they comply with and implement good SHE Practices.

There are three **types of safety audit**: compliance **audit**, program **audit**, and management system **audit**.

What is the difference between an audit and inspection?

At a high level, **inspections** are a "do" and **audits** are a "check". An **inspection** is typically something that a site is required to do by a compliance obligation. An **audit** is the process of checking that compliance obligations have been met and required **inspections** have been done

How safety audit is conducted?

The basics of a **safety audit:**

First determine the organizations safety requirements and expectations, be they regulatory, program, organizational or cultural. Remember, **auditing** is not a one-time process. Each type of **audit** should be **conducted** at least annually but it helps to know what's going on between the **audits**

What is safety audit HSE?

Safety audit is a planned, independent, documented and a systematic approach of determining the success level of the Health and **Safety** Management System. It helps to identify emerging **safety** issues before they become problems and also serve as a catalyst for necessary changes to improve employee **safety**.

The Audit Process

- ❏ Step 1: Define Audit Objectives- Prior to the audit, AMAS conducts a preliminary planning and information gathering phase. ...
- ❏ Step 2: Audit Announcement. ...
- ❏ Step 3: Audit Entrance Meeting. ...
- ❏ Step 4: Fieldwork. ...
- ❏ Step 5: Reviewing and Communicating Results. ...
- ❏ Step 6: Audit Exit Meeting. ...
- ❏ Step 7: Audit Report.

Audit is an expert assessment of an organisation's health and safety policies, systems and procedures.

"Are we legally obliged to conduct a health and safety audit?" is the first thing many people want to know. The short answer is "No". But a longer and more helpful answer is "You're not legally obliged to, but it's highly recommended."

18: HSE Audit:

Regular HSE audits are widely recognised as best practice for companies of every type and size. Much more than just a box ticking exercise or defensive measure, they can bring many positive benefits to the business.

Primarily, audits enable you to protect employees and customers from harm. But ultimately a correctly conducted audit can also safeguard the very existence of the company, because a serious health and safety incident resulting in prosecution can bring severe reputational damage and massive financial penalties.

The audit in detail

A typical health and safety audit will cover:

- Documentation: checking that you have suitable health and safety policies, process documents and suitable arrangements for harmful substances in place
- Interviews with managers, heads of departments, and 'shop floor' personnel (who often have invaluable insights based on everyday working experience)
- Checking whether policies are being adhered to

In-house or outsourced?

Health and safety audits can be internal or external, conducted by a member of your staff, or by external independent consultants.

Internal

To conduct an internal audit, you require at least one team member with the right training and qualifications – a NEBOSH Certificate. The main advantage of the in-house option is greater control over the process. It can take place in stages.

External

An external health and safety audit is an "independent assessment of your safety management systems".

Structured Process

There are different elements of an HSE audit, depending on the exact nature and how comprehensive the specific audit is to be. The primary elements are:

- HSE Management Audit
- HSE Site Compliance Audit
- HSE Plan Audit
- Process HSE Audit
- Product HSE Audit

An individual HSE audit can contain one or more of these elements, with some elements being required at different times.

HSE Management and Site Compliance Audit

These are often combined by their very nature and as one can indicate overall compliance within the structure of the other. Preparation for this type of scenario requires on-going HSE processes to be implemented and regularly evaluated and that all compliance procedures be followed up as quickly as possible. Deficiencies discovered during this type of audit can automatically trigger further audits.

HSE Plan Audit

HSE management audit, reviews the current plans of actions concerning all aspects of the HSE system. This includes contingency planning, emergency planning, and

18: HSE Audit

the effectiveness of any sub planning units. In addition, this can include practice scenarios to ensure that the current plans are efficient and effective. Preparation for this type of audit is simply a case of making sure all the appropriate documentation and personnel are in position are working as intended.

Process and Product HSE Audit

These two audits look at the specific process safety and the final product to ensure that they meet all the HSE standards, as set out by management and they fulfil all objectives in regards to HSE compliance. In some industries Product HSE audits are not necessary; however, Process HSE audits are required by almost all. Preparation in regards to this audit is limited as there should be on-going compliance with HSE objectives at all times.

Importantly, HSE audits should be conducted by independent, outside sources to ensure there is little to no attempts to derail the audit. Given that the audit is to ensure the proper health, safety and environmental compliance with the numerous rules and regulations therein, it is better to partner with a knowledgeable and experienced HSE audit provider that can not only conduct the audit, but provide recommendations for improvements based on the audit. By doing so, your business is not only prepared for the current audit, it is in a position to better succeed with the next one as well.

Are there many types of safety audit?

Well, roughly speaking, we could sum up at least six types of audits which are related to health and safety:

Health and safety audits

Objectives of this type of safety auditing are to inform the company:

- How well it is performing in H&S
- Whether managers and others are meeting the standards which the company has set itself
- Whether the company is complying with the H&S laws which affect its business With the view that the company making any improvements identified as necessary from this information.

Audit of a Health and Safety Plan

Audit of a Health and Safety Plan is the review of this plan at the end of the year.

The audit consists of two parts. The first is intended to provide a simple overview of progress in terms of time. The second is intended to expand on the information provided in the first part by giving reasons as to why any missed deadlines were not met, detailing any benefits gained by the activities undertaken in the time period covered by the plan, and including any other relevant information that will assist in drawing up the plan for the next 12 months.

The audit looks into the progress of the existing plan as well as the contents and format of the plan.

"Walk around audit"

A "walk around audit" is to determine whether the health and safety policies of the company are being properly implemented and to identify areas in which policy effectiveness needs to be improved.

18: HSE Audit:

H&S management audits

One of the main problems with H&S audits of the above type is that they tend to examine H&S problems from the symptoms rather than causes of the problems. They rarely focus entirely on the management of H&S.

H&S Management audits look into the following areas:

1. Does the company have adequate procedures for identifying specific H&S requirements which apply to its organization?
2. Are the procedures followed and are the responsibilities set out clearly and understood?
3. Does the company's H&S policy documentation include adequate procedures for identifying hazards which exists at the workplace, and for assessing regularly the risks to employees and others affected by the workplace and workplace activities in order to identify the measures needed to avoid their exposure to risks of harm?
4. Are adequate risk assessment procedures also set out for hazards of products and /or services supplied by the company in order to identify the measures needed to avoid risks of harm to people such as distributors, customers, end-users and members of the public?
5. Are the procedures in point 3. and 4. followed, and are responsibilities set out clearly and understood?
6. Does the company have adequate procedures for setting, reviewing and revising as necessary its health and safety standards for meeting specific H&S requirements and for meeting its general duties to protect employees and others form risks identified in the company's risk assessments?
7. Do the procedures for setting company standards include the identification of measurable targets

which can be audited to monitor the level of compliance with company standards?
8. Are the procedures in 5. and 6. followed and are responsibilities set out clearly and understood?
9. Does the company have adequate procedures for planning, implementing, controlling, monitoring and reviewing the measures identified in 3. and 4. ?
10. Does the company have adequate procedures for carrying out H&S audits to check that the procedures in 9. are followed and that the measures in 3. and 4. are effective?

Project Health, Safety & Environmental (HSE) auditing

Project HSE auditing provides the method for monitoring and controlling HSE activities and procedures throughout the life of the project. It comprises of two activities:

- Formal audits
- Regular and ad-hoc inspections

The formal audits provide a more comprehensive and formal assessment of compliance with HSE procedures and plans. They should be carried out at key points during the project life cycle.

PROCESS SAFETY AUDITS

It is a self-evaluation audit which aims at:

1. gathering all relevant documentation covering process safety management requirements at a specific facility
2. determining the program's implementation and effectiveness by following up on their application to one or more selected processes.

18: HSE Audit:

Process Safety Audit is important in the product design and development stages. It is to ensure that the company had adequately protected the user of a product from hazards that it did not know existed. This type of audit is to

- Identify and classify hazards associated with the product i.e. catastrophic, critical, occasional, remote, or improbable
- Develop a hazard risk index and priority setting
- Get employee to present design alternatives and to review for feasibility

Safety Management Audit Rating Tool (SMART)

Safety Management Audit Rating Tool (SMART) is developed by the Society of Accredited Safety Auditors for assessing the occupational safety and health management system of contractors in construction work.

The design of SMART is based on British Standard BS8800:1996 Guide to Occupational health and safety management systems and standards set by local legislation. SMART provides an easy step by step assessment of the site health and safety management functions and its compliance with local legislation. Its aim is to unveil deficiency and weakness of the system for the management which in turn would help the management to further improve on site health and safety management.

SMART can be used as an in-house management tool for self-assessment and it can be used by an external auditor for an independent audit. A scoring system is introduced to help in setting baseline for further improvement and for comparison among sites of the company.

INDUSTRY EXAMPLES OF PROCESS SAFETY AUDITS

Example 1:

Chemical Industry - Process Safety Audit

Step 1: Preparation

- Identify Audit Scope: Define the boundaries and objectives of the process safety audit.
- Assemble Audit Team: Form a multidisciplinary team with expertise in process safety, engineering, operations, and regulatory compliance.
- Review Documentation: Examine existing process safety documentation, including procedures, risk assessments, and incident reports.

Step 2: Site Visit and Observation

- Walkthrough: Physically inspect the facility, process units, and equipment to observe operations and identify potential hazards.
- Interviews: Conduct interviews with personnel at various levels to gather information on safety practices, training, and awareness.
- Data Collection: Collect data on process parameters, equipment conditions, and safety measures.

Step 3: Document Review

- Analyze Procedures: Evaluate the adequacy and adherence to process safety procedures, including standard operating procedures (SOPs) and emergency response plans.

18: HSE Audit:

- Review Risk Assessments: Examine Hazard and Operability Studies (HAZOP), Process Hazard Analyses (PHA), and other risk assessments to ensure completeness and effectiveness.
- Compliance Check: Verify compliance with regulatory requirements and industry standards.

Step 4: Systems and Controls

- Evaluate Safety Systems: Assess the effectiveness of safety instrumented systems (SIS), emergency shutdown systems, and other safety controls.
- Review Maintenance Practices: Examine maintenance procedures for critical safety equipment and verify their compliance.

Step 5: Management Systems

- Management Review: Evaluate the commitment of management to process safety, including their involvement, communication, and resource allocation.
- Training and Competency: Assess the training programs and competency of personnel involved in process safety.

Step 6: Documentation of Findings

- Compile Audit Findings: Document all observations, findings, and recommendations.
- Prioritize Findings: Prioritize the findings based on risk and significance to process safety.
- Draft Audit Report: Prepare a comprehensive audit report detailing the findings, recommendations, and suggested corrective actions.

Step 7: Presentation and Review

- Present Findings: Conduct a presentation to relevant stakeholders, including management and process safety teams, to discuss the audit findings.
- Obtain Feedback: Solicit feedback and input from stakeholders to ensure a thorough understanding of the findings.

Step 8: Corrective Actions

- Develop Action Plan: Collaborate with the organization to develop an action plan for addressing the identified deficiencies and implementing corrective actions.
- Monitor Progress: Regularly monitor and track the progress of corrective actions to ensure timely implementation.
- Continuous Improvement: Encourage a culture of continuous improvement, learning from audit findings to enhance overall process safety.

Example 2:

Petrochemical Industry - Process Safety Audit

The steps for a Petrochemical Industry Process Safety Audit would align with the ones mentioned above, with specific considerations for petrochemical processes, such as:

- **Hazardous Chemicals Handling:** Emphasis on the safe handling, storage, and transportation of hazardous chemicals commonly used in petrochemical processes.

18: HSE Audit:

- **Emergency Response Preparedness:** Evaluation of emergency response plans, including procedures for dealing with chemical spills, fires, and other critical incidents.
- **Mechanical Integrity:** Rigorous inspection and assessment of equipment integrity, corrosion protection, and preventive maintenance programs for high-pressure vessels and piping systems.
- **Process Hazard Analysis for Unique Petrochemical Processes:** Special attention to specific petrochemical processes and their unique risks, such as cracking units, reformers, and distillation columns.
- **Security Measures:** Consideration of security measures to safeguard against potential threats or sabotage, including access control and monitoring systems.

Each industry-specific audit would tailor these steps to address the distinct challenges and risks associated with chemical and petrochemical processes, ensuring a thorough evaluation of process safety.

Chapter 19

LEADING & LAGGING INDICATORS OF SAFETY

One way to improve the effectiveness of your safety process is to change the way it is measured.

Measurement is an important part of any management process and forms the basis for continuous improvement. Measuring safety performance is no different and effectively doing so will compound the success of your improvement efforts.

Finding the perfect measure of safety is a difficult task. What you want is to measure both the bottom-line results of safety as well as how well your facility is doing at preventing accidents and incidents. To do this, you will use a combination of lagging and leading indicators of safety performance.

Lagging indicators of safety performance
What is a lagging indicator?
Lagging indicators measure a company's incidents in the form of past accident statistics.

Examples include:

Injury frequency and severity

- Lost workdays
- Worker's compensation costs
- OSHA recordable injuries

Why use lagging indicators?

Lagging indicators are the traditional safety metrics used to indicate progress toward compliance with safety rules. These are the bottom-line numbers that evaluate the overall effectiveness of safety at your facility. They tell you how many people got hurt and how badly.

The drawbacks of lagging indicators:

The major drawback to only using lagging indicators of safety performance is that they tell you how many people got hurt and how badly, but not how well your company is doing at preventing incidents and accidents.

The reactionary nature of lagging indicators makes them a poor gauge of prevention. For example, when managers see a low injury rate, they may become complacent and put safety on the bottom of their to-do list, when in fact, there are numerous risk factors present in the workplace that will contribute to future injuries.

Leading indicators of safety performance

What is a leading indicator?

A leading indicator is a measure preceding or indicating a future event used to drive and measure activities carried out to prevent and control injury.

Examples of leading indicators include:

- Safety training
- Reduction of MSD risk factors

19: Leading & Lagging Indicators of Safety:

- Employee perception surveys
- Safety audits
- Ergonomic opportunities, identified & corrected

Why use leading indicators?

Leading indicators are focused on future safety performance and continuous improvement. These measures are proactive in nature and report what employees are doing on a regular basis to prevent injuries.

Best practices for using leading indicators

Companies dedicated to safety excellence are shifting their focus to using leading indicators to drive continuous improvement. Lagging indicators measure failure; leading indicators measure performance, and that's what we're after!

Leading indicators should:

1. Allow you to see small improvements in performance

2. Measure the positive: what people are doing versus failing to do

3. Enable frequent feedback to all stakeholders

4. Be credible to performers

5. Be predictive

6. Increase constructive problem solving around safety

7. Make it clear what needs to be done to get better

8. Track Impact versus Intention

While there is no perfect or "one size fits all" measure for safety, following these criteria will help you track impactful leading indicators.

Conclusion

To improve the safety performance of your facility, you should use a combination of leading and lagging indicators.

When using leading indicators, it's important to make your metrics based on *impact*. For example, don't just track the number and attendance of safety meetings and training sessions – measure the impact of the safety meeting by determining the number of people who met the key learning objectives of the meeting / training.

Chapter 20

SALARY NEGOTIATION

Although negotiating salary, may be common in some places it is NOT recommended in general. It may give the potential employer a negative impression about the candidate.

Alternately the following discussions could be more fruitful if needed: -

"I am excited by the opportunity to work together."

Many people often think of salary negotiation as a battle: you, trying to get as much as you possibly can, versus your employer, trying to stay within budget. However, this type of thinking can be counterproductive.

Unless you know for sure that you are indispensable, and very few of us ever are, successful negotiation should never become adversarial. That is a bad sign that the process has broken down or it will.

"Based on my research

You can get a higher salary than the one that you were offered, but it needs to be grounded in reality. Rather than just throwing out a number that you think sounds nice, you need to do your homework on what your skills are worth in order to provide a compelling case for your employer to compensate you accordingly.

"Is that number flexible?"

If an employer offers a number that's below your desired range, pushing back is essential but you want to make sure you handle it with tact. Saying "is that number flexible at all" is a graceful way to "[give] the employer the opportunity to offer more, or even mention other perks you might be able to gain if a higher salary isn't in the picture

"I would be more comfortable if..."

Blunt phrases like "I need" or "I want" can be a turn-off to employers. But expressing your desired salary with this phrase "is a better way to let the recruiter or hiring manager know specifically what you're looking for so they can focus on that dimension of your job offer"

"If you can do that, I'm on board."

Most of the times, recruiters are just as anxious as you for salary negotiations to come to a close. So, if you can specifically spell out what it would take for you to accept an offer, you'll be doing recruiters and hiring managers a favour.

"When you get to this phase of the negotiation, you want to make it clear to the recruiter or hiring manager that saying 'Yes' will end the negotiation so they're more comfortable

20: Salary Negotiation:

"Do you mind if I take a couple of days to consider your offer?"

Even if a job offer exceeds your expectations, try to play it cool. "The first thing you should do when you receive a job offer is ask for time to consider it. "This little phrase accomplishes several things. First of all, it buys you time to consider the offer, determine the appropriate counteroffer. [But] it also enables you to move the negotiation to email if it's not already there.

Your salary negotiation will be more successful if you carefully choose your counteroffer amount and clearly explain why you're worth it.

"Thank you."

"At the end of the salary discussion, be sure to thank the person for taking the time to sit down with you, just to maintain your professionalism.

An employer is much more likely to accommodate the wants and needs of somebody that shows them respect.

Chapter 21

INTERVIEWER'S TEN FAVOURITE QUESTIONS

Interviews are one of the most challenging parts of trying to get a new job.

When you're selling yourself and your skill set, you need to have just the right answer for everything.

When you aren't sure what interview questions you'll be asked, it can be difficult to prepare.

Generally, many interviewers ask the same or similar questions as given below:

Question 1: Tell me about yourself?

Answer: This surely is the number 1 Interview Question, that candidates from across the world come across. It's the same in US/UK/India /Dubai/Gulf countries. Talking about 'things not related to work' is a big NO-NO

My advice to you is to prepare this answer beforehand. Talk about some responsibility that you enjoy doing, especially one that is related to the job position you are interviewing for.

Question 2. Why did you leave your last job?

Answer: This might be one of the trickiest ones- Always answer in the positive. Never ever give answers that will show you in a bad light. If you've had problems with your manager, or a colleague etc. DO NOT say that. Instead say that you wanted a challenge in life, a better opportunity etc.

Question 3. Are you applying for other jobs?

Answer:

Honesty pays great rewards. If you have applied for similar jobs with competitor organizations, then say so. But do not divulge too much detail. Also, if you've applied for different types of jobs, then it's advisable to keep quiet.

Question 4. Why do you want to work for this organization?

Answer:

21: Interviewer's Ten Favourite Questions :

This is another tricky question. Researches about the organization before you go for the interview. This will give you some idea about how you could fit into the bigger picture. Be honest & sincere. Talk about your long-term goals and how the work that you do here will help you achieve them.

Question 5. What is your Expected Salary?
Answer:

This is where most people fumble. It's advisable to not speak out a figure. Instead, say that the salary package depends upon what exactly this job entails.

Then ask the interviewer what salary range the company has, for this job position. If the interviewer quotes a salary range, that's good, and if he does not, you've already answered your part.

Question 6. How long would you expect to work for us if hired?

Answer:

Do not give a figure for this one too. What they are looking for is stability from employees. Hence an answer like, "I'd like to work for a long time in this organization", or, "I'd like to work here as long as I can provide value and growth to this organization", should be good answers.

Question 7. Explain how you would be an asset to this organization?

Answer:

This is one question that you should be waiting for. Talk honestly and positively on this one highlighting your strongest skills, especially suited to the job position at hand.

Question 8. Why should we hire you?

Answer: Stress upon the fact that your skills match those that the organization is looking for.

Question 9. What is your greatest strength?

Answer: You can give various answers for this one. Like for example, Team player, Work and deliver under pressure, leadership skills, never say die attitude etc. Remember not to say big, fabulous sounding words, else it will sound quite like lies.

Question 10. What have you learned from mistakes on the job?

Answer: "Humans make mistakes. I have made some too. For example, I thought I could do it all alone. Then, I realized that 1 and 1 make Eleven not 2." Positive examples lie these are sure to impress the interviewer

Chapter 22

COMMON QUESTIONS

What inputs are required by FSES in order to conduct the PHA Study?

As a minimum the following information would be required in order to conduct the workshop:

- Existing PHA / HAZOP report (if available)
- P&ID's
- Cause and Effects Diagrams
- Facilities Design
- Operating Data and Procedures
- Maintenance Data and Procedures
- Interlock List
- Equipment Data Sheet

Based on the project requirements, additional information may be required, which will be highlighted within the Terms of Reference (ToR).

Who is required to attend the PHA Workshop?

As a minimum the following personnel would be required in order to conduct the workshop:

- Process Engineer
- Controls and Instrumentation Engineer
- Process Safety Engineer
- Maintenance representative
- Operations representative

Based on the project requirements, additional personnel may be required to attend the workshop, which will be highlighted within the ToR.

What is the expected output of the PHA Study?

On award of the study FSES will issue a project ToR, which will highlight the assumptions that shall be made in the study, along with the workshop details, methodology and data sources that will be utilised as well as any further information required from the client.

Upon acceptance of the ToR, FSES will facilitate the PHA study through a workshop providing the facilitator and scribe. Once the workshop has been conducted FSES will prepare a PHA report describing the facility, the scope of work, a detailed methodology, the PHA worksheets, a summary of the PHA actions and any recommendations based on the discussions during the workshop. FSES highly recommend that following on from the PHA study, a SIL

Determination analysis is conducted in order to determine the SIL requirements for the SIF are identified during the workshop.

22: Common Questions :

'FIRE TRIANGLE'

A fire triangle represents the three elements, which causes a fire in a combustible mixture. The three elements are fuel, air and ignition source.

Question 1. What is Personal Protective Equipment?

Answer: It is equipment used to protect the person from hazards such as dust, dirt, and fumes and sparks etc. It is the barrier between hazard and person.

Question 2. What is Safety Audit?

Answer: Safety audit is the process that identifies unsafe conditions and unsafe acts of the plant and recommends safety improvement. It evaluates the unsafe condition noticeable to naked eye during work through the plant (Stores, civil work, erection work). It evaluates the safety factors in the plant on the base of engineering, analysis, testing, and measurement.

Question 3. What is Safety Tag?

Answer: Safety tag can be defined as a tag made of card board or paper board on which warning/safety instructions are written for employees.

Question 4. What is Manual Handing?

Answer: The process of lifting, carrying and stacking materials by men physically is called manual handing.

Question 5. What is Safety Policy?

Answer: Any company has a social and legal obligation to provide a safe and healthy working environment to all its employees.

The safety and health of all employees is one of prime concerns of the company.

Every company will be requiring the policy both in letter and in spirit.

The company shall comply straight with act, laws, rules and regulations

The company shall impart Training in health safety and occupational health to all employees.

The company will adopt own safety and health standards where laws may not be available.

Question 6. What are Vehicles / Mobile Permit?

Answer: The permit is required for taking any vehicle or mobile equipment having a diesel or petrol operated engine into hazardous area.

Question 7. Explain what is Earthling?

Answer: Earthling means connecting the equipment in use to the general mass of the earth through conducting line.

22: Common Questions :

Question 8. What is Inspection?

Answer: Inspection is finding out hazards according to checklist prepared with reference to the department/operations by the people who are familiar with the plant.

Question 9. What are the causes of accidents in Manual Handling?

Answer:
1. Improper lifting
2. Carrying too heavy loads
3. Improper gripping
4. Failure to use PPE
5. Lifting greasy, oily and irregular objects
6. Poor physique

Question 10. What is a typical work day for a Safety/HSE Supervisor?

Answer: At the beginning of each day

Inspect the work site to make sure that it is hazard-free.

Once the work site is secured, verify that all tools and equipment are adequate in supply.

As soon as the work orders are delivered, provide workers with security guidelines and carry out drills if needed.

During the workday, monitor workers to ensure that they are working according to the enforced safety policies and that any problems or accidents are quickly addressed.

Question 11. What are the Hazards encountered in Manual Handing?

Answer: During manual handling the following hazards are most likely:
1. Cutting fingers due to sharp edges
2. Burns due to handing of hot articles

3. Foot injuries due to dropped articles
4. Slipped disc due to improper posture in lifting on object
5. Strains to wrist or fingers
6. Sprains, wounds hernias, fractures

Question 12. Explain your experience of handling Safety Documentation?

Answer: I have worked extensively with management teams to develop new safety policies, augment existing ones and handle draft construction related specifications. I have written safety proposals and conducted safety programs as well.

Question 13. What is an Accident?

Answer: It is an unexpected or unplanned event which may or may not result in injury or damage or property loss or death.

Question 14. What is Work Permit?

Answer: Work permit is a "written document" for permission to undertake a job by area in charge or it is written document issued by the area in charge to the performer to undertake the specific job.

Question 15. What are the precautions to be taken before starting a Welding job?

Answer:
- Remove all combustion material from the place of welding
- Clear the work area and cover wooden floor with fire proof mats.
- Welding machine should be kept within the visibility of the welders.
- Erect fire resistance screen around the work place
- All welding cables should be fully insulated

22: Common Questions :

- All welding machines shall be double earthed
- Welding area should be dry and free from water
- Keep the fire extinguisher / sand ready
- Use leather hand gloves, goggles and helmets
- Switch off the power when welding is stopped
- Do not permit the helper to do welding
- Do not shift the welding cable unless the electric power is switched off.
- Terminal of the welding cables should have 3-cable with lugs and kept tight.
- Oxygen hose in black and Acetylene hose in red colour as per standard
- NRV of the blow torches should be maintained properly to avoid back fire
- Welders should be trained properly
- Cylinders should be stored in a cold dry place away from heat and direct sunlight.
- Proper housekeeping, good ventilation in the working area
- Smoking should be avoided in the welding area
- Hose connection should be made properly
- Barricade the work area and put a sign board
- Rolling of cylinders should be avoided
- Flash back arrestor should be attached in each cylinder
- Any leaky cylinder should be kept separately

Question 16. What are the Safety Rules to be followed in using Compressed Air?

Answer:

- Only authorized persons should use compressed air.
- The body or clothes should not be cleaned with compressed air.
- Compressed air hose pipes should not be placed across passage ways
- Leakage of compressed air should not be tested with hands.
- While working with tools run by compressed air safety shoes to be used.
- The tools should not be kept on position when not in use.

Question 17. What are the precautions to be taken during Excavation?

Answer:

- Excavation area should be suitably barricaded
- Put sign boards, lights and flags
- Avoid heavy vehicle coming near the sides
- PPE like helmet, safety shoes should be used
- Keep the excavated soil far (at least 5 feet distance)
- Excavated sides should be sloped back to a safe angle

22: Common Questions :

- Hand excavation should be done at the places where UG pipes or cables are present
- Cutting shall be done from top to bottom
- All narrow trenches 4 feet or more deep shall be supplied with at least one ladder
- While excavating on the slope whose height is over 10 feet men should use safety belts.

Question 18. What is Safety Management?

Answer: Safety management is the art and science of planning, administration and improving various functions to achieve the safety objectives.

Question 19. How do we Control the Chemical Hazards?

Answer: The chemical hazards are controlled by engineering methods, administrative methods and PPE.

Question 20. What is housekeeping?

Answer: Housekeeping means not only cleanliness but also orderly arrangement of operations, tools, equipment storage facilities and suppliers.

Question 21. What is working at height?

Answer: Any work above 2 meters from ground is called work at height. You are working at height if you: work above ground/floor level or could fall from an edge or through an opening or fragile surface.

Question 22. Explain Safety in using Hand Trucks?
Answer:
- The truck should be inspected
- The axles should be greased well
- Safety shoes should be worn by the operators.
- The load should be balanced and the weight of the load should not fall on the axle
- The hand cart should not be wider than the width of the hand truck.
- The hand cart should be pushed and not pulled
- The truck should not be placed on path ways.

Question 23. What is delegated Work Permit?
Answer: Delegated work permit is used for areas requiring light control for example in Fabrication, yards. It is valid for 30 days.

Question 24. What are the main provisions in "The Factory Act"?
Answer: Health, safety, welfare, hours of work, employment, person, occupational disease, special provision and penalties and procedures.

Question 25. Explain the Control Measures of Radiography?
Answer:
- Barricade the area
- Remove all un-necessary persons away from site

22: Common Questions :

- Check radiation level with dosimeter
- Use lead shields
- Put a sign board
- Risk - tissue damaged
- Use special filter glass
- Use lead coated aprons

Question 26. Explain what is Ingestion?

Answer: Entry of harmful materials through mouth is called ingestion.

Question 27. What Is Hot Work Permit?

Answer: It is a permit issued for work which involves spark flame, or temperature.

Question 28. What are the various types Work Permit?

Answer:

There are two types of work permit

- Cold work permit
- Hot work permit

The hot work permit is further classified into 3 types:

- Normal hot work permit
- Blanket hot work permit
- Delegate hot work permit

Question 29. What Is Industrial Hygiene?

Answer: Industrial hygiene is defined as the art and science of the presentation and improvement of the health and comfort of workers.

Question 30. Explain various types of Sign Boards?
Answer:

- Mandatory
- Information
- Fire or explosion
- Caution
- Warning

Question 31. What is First Aid?
Answer: First aid is temporary and immediate care given to the victim of an accident.

Question 32. What is Accident Prevention?
Answer: Accident prevention may be defined as an integrated program directed to control unsafe mechanical or physical condition.

Question 33. Explain different types of Hazards?
Answer:

- Mechanical hazards - inadequately guarded machines parts
- Chemical hazards - of toxic gases, vapours, fumes, smoke & dust.
- Electrical hazards: inadequately insulated live wires
- Fire hazards – chemical reaction, electrical arcs

22: Common Questions :

- Radiation hazards – dazzling light, infrared rays & ultra violet rays
- Pollution – air, water & noise pollution

Question 34. Explain the general Safety precautions in Construction?

Answer:

- Adequate first aid equipment should be kept ready
- Adequate fire fighting equipment should be available
- All general electrical rules should be followed
- Sufficient lighting arrangements should be available at night for work
- Workers at height should wear safety belts (harness)
- Work men handling cement should be provided with goggles, rubber gloves nose mask and rubber boots.
- The moving parts of grinding machines used construction site should be covered with guards

- Excavated material should not be kept near the excavation
- For very short duration of work red flags must be hoisted and for more duration red banners must be stretched
- Defective tools should not be used
- The worker should not carry tools in his hands when climbing a ladder
- Excavation should be guarded by suitable fencing

Question 35. What is a Grinder?

Answer: A grinder is a power tool or machine used for grinding; it is a type of machining using an abrasive wheel as the cutting tool.

Question 36. What is Welding?

Answer: The process of joining of metals either by electrical process or by gas is called welding.

Question 37. What are the duties of a Process Safety Engineer?

Answer: Apart from enforcing key safety standards, it is important to monitor manpower, conduct regular site inspections, record all violations and conduct safety trainings.

22: Common Questions :

Question 38. What is Noise?

Answer: Any unwanted sound which causes irritation to the ears caused by mechanical movement.

Question 39. What is a Safety Survey?

Answer: Safety surveys are conducted to have detailed observations of all types of unsafe physical and environment conditions as well as unsafe practices. These surveys are committed to the health, safety and comfort of the workers.

Question 40. What is the responsibility of Workers towards maintaining a Safe working place?

Answer:

Report unsafe condition to supervisor

Do not operate the machines without knowing the operation. Follow the SOP (Standard operating procedure)

Before starting the machine, check whether the machine is in good working condition or not

Use correct tools

Follow the safety rules

Do not horse play

Question 41. Explain the usage of Crane and Slings?

Answer:

Only authorized and competent person should operate cranes

The correct sling must be used for the load to be lifted

Lifting equipment must be certified from competent authority and mark with its SWL

Never use Crane for loads in excess of its SWL

Cables and slings must be padded when passing over sharp edges of equipment

Check the condition of the ground before parking the crane

All moving parts must be guarded

Uncertified chains, ropes, slings and hooks should not be used

All slings to be inspected and certified by third party inspectors

Never stand or work under a suspended load

Place the out riggers on firm ground

Guided ropes shall be used to control swing of lifted material

Never operate the crane at the time of speed wing

Lifting over live equipment should not be encouraged

The crane should undergo periodical maintenance as per manufactures recommendations

Question 42. Explain safety precaution for Scaffold?

Answer:

- Wooden board should not be painted
- Wooden board should not have any cracks
- Check for rust in pipes / clamps
- Clamps should be fixed and be of good quality

22: Common Questions :

- Board thickness should be 3.4 cms without any bends
- The construction must be rigid
- Use of good quality material
- The wooden bellies should not have joints
- Vertical poles should be less than 6 feet
- Chains and ropes should be used for the suspension of scaffoldings
- Never throw any materials from height
- Use safety harness while working at height above 6 feet

Question 43. What are the duties of a Process Safety Engineer?

Answer: The duty of Process Safety Engineer includes the following: -

Prepare tool box talk

Prepare monthly statistics

Prepare checklist

Prepare Accident reports

Attend Management meetings

Arrange the safety classes/training

Arrange monthly safety bulletin

Inspection of fire extinguisher

Arrange first aid training classes

Arrange safety competitions like quiz, slogan, poster competitions exhibition etc.

Question 44. What is UEL?

Answer: Upper Explosive Limit (UEL) - The maximum proportion of vapour, gases and dust in air above which the flame does not occur on contact with a source of ignition is called UEL.

Question 45. What is Excavation?

Answer: Making a hole or tunnel by digging the ground by man or machine is called excavation.

Chapter 23

GENERAL SAFETY QUESTIONS

Question 1. When should you change the Battery of Smoke Detector?

Answer: It is recommended that the battery be changed at least once a year or whenever the detector makes a chirping sound, indicating low battery power.

Question 2. What do dry Sprinkler Systems contain?

Answer: Compressed Air or Nitrogen.

Question 3. Name two types of Fire Alarm Systems?

Answer: The alarm systems are either manual or automatic.

Question 4. Are all "Climbing Walls" required to be inspected?

Answer: All climbing walls over 10 feet in height are required to pass an annual safety inspection and have a state operating permit. Climbing walls less than 10 feet in height are exempt from the inspection and permitting process.

Question 5. What two requirements are necessary for Escape in case of Fire?

Answer:

The sound of the alarm.

The means of escape and route to follow.

Question 6. What two Colors are generally accepted for Danger and Non-danger?

Answer: Red for danger and Green for non-danger.

Question 7. What is the color of a Fire Exit sign?

Answer: White symbols on a green background.

23: General Safety Questions:

Question 8. What type of Extinguisher would you use on Petroleum?

Answer: Dry Chemical Powder (DCP) or Foam.

Question 9. What action should you take in case you observe a fire at workplace?

Answer:

Raise the alarm,

Get everyone out,

Attack the fire if you are competent and are aware of Fire extinguisher working

Question 10. Name four types of Fire Extinguishing Agents?

Answer: Water, Foam, CO2, Dry Chemical Powder.

Question 11. Is it advisable to grease the valve of Gas cylinders?

Answer: It is not advisable to grease the valve because Oil and Grease, can react chemically causing an explosion.

Question 12. Name the two types of Smoke Detectors?

Answer: Ionisation Detectors and Optical Detectors.

Question 13. How should we dispose of Flammable waste in a workroom?

Answer: Put it in a closed metal bin and empty regularly.

Question 14. What should you do if a person's clothes catch fire?

Answer: Roll them in a blanket or use a non-asphyxiate extinguisher.

Question 15. Why are oily or paint covered rags dangerous?

Answer: They may be liable to spontaneous combustion.

Question 16. Name the Fire Extinguishing Agents recommended for use on Electrical Fires?

Answer:

- CO_2
- Dry Powder.
- Sand.

Question 17. How do water type extinguishers, extinguish Fires?

Answer: By cooling the burning material to a point where it will not burn.

Question 18. How does DCP Extinguishers, Extinguish Fires?

Answer: By stopping chain flame reaction and smothering.

Question 19. What type of Extinguisher is best suited for free-flowing liquid fires?

Answer: Dry Chemical Powder (DCP).

23: General Safety Questions:

Question 20. Name the means by which heat may be transmitted and result in spread of Fire?

Answer: Conduction, Convection and Radiation.

Question 21. Give some sources Of Fire?

Answer:

- Sparks
- Flames.
- Hot Surfaces.
- Radiant Heat.

Question 22. What are the characteristics of flammable substances that are crucial in creating an Explosive Atmosphere?

Answer:

- The density relevant to air.
- The flash point.
- The ignition temperature.
- The boiling point.
- The upper and lower explosive limits.

Question 23. What is the expanded form of BCF?

Answer: The expanded form of B.C.F. is Bromo chloro-difluo-methane; also called Halon 1211. It is used as Fire suppressant.

Question 24. What happens when UV Detectors detect a Fire?

Answer: The UV detectors create annunciation, audible alarm on the control panel and siren in the field.

Question 25. What is the Time Delay between Fire Sensing by a UV detector and confirming with an alarm?

Answer: Generally, it is set for 4 secs. The UV detectors initiates a fire alarm only when the UV is detecting the fire continuously for 4 secs

Question 26. What happens when a Heat Switch actuates?

Answer: On detection of heat, the heat switch initiates the following:

Annunciation, audible alarm on the control panel, a siren in the field, shut down of the equipment and release of fire extinguisher.

Question 27. Explain the operating Principle of a Gas Monitor?

Answer: Gas monitor measures the imbalance in the current loop caused by its active and inactive filaments in the presence of a combustible gas.

Question 28. Explain the Calibration procedure of a Gas Monitor System?

Answer: Gas sensor loop current or voltage at the sensor head is set as per the manufacturer's recommended value.

Gas monitor zero is adjusted to the instrument air.

A test gas with a known quantity of combustion gas (Generally methane 2% by volume) is fed to the sensor and the span is adjusted to read 40% on the monitor scale

23: General Safety Questions:

The calibration procedure is repeated until the zero and span reads correctly.

Question 29. What are the "Alarm" and "Shut Down" settings on a Gas Monitor?

Answer: Generally, on the gas monitor, the alarm is set at 20% rising and the shutdown is set at 60% rising

Chapter 24

MISCELLANEOUS - QUESTIONS

An insight into the following questions & answers will help in better understanding of widely used basic safety terminology.

Question 1. What is safety?

Safety is ensuring that people are not exposed to situations that can readily result to accidents or any kind of harm. This means that in workplaces where safety is enhanced, there are measures that are put in place to make sure that the people working there are not exposed to conditions that can lead to the people being hurt or harmed.

Question 2. What is accident?

Answer: An accident is a situation which happens unexpectedly and causes harm. Accidents usually occur without the people knowing. Majority of accidents happens because of negligence. Accidents can also happen because of doing things in the wrong ways instead of doing them the way they are supposed to be done.

Question 3. What is hazard?

Answer: Hazard is something that has a probability of causing harm if not handled properly. This means that there is a risk of certain kind of harm if the necessary safety precautions are not implemented properly. The main aspect of hazard is that if everything is done properly the risk of harm is significantly reduced.

Question 4. What is risk?

Answer: Risk is the increased chances of something happening. In HSE risk involves probability of accidents and incidents happening in workplaces. High risk means that there are higher chances of an accident or incident happening. This is the primary reason why HSE advices for reduction of workplaces risks to ensure chances of accidents and incident are reduced.

Question 5. What is safety policy?

Answer: A safety policy defines guidelines put in place to enhance safety. The policy outlines the things that need to be done and the things that need to be avoided so as to avoid accidents and incidents. The policy also stipulates measures that need to be put in place to provide protection against things that can cause harm.

Question 6. What is incident?

Answer: An incident is when something that is out of the normal happens. This does not necessarily have to cause harm but it deviates from the normal procedures. According to the HSE department, most incidents usually result into accidents and this can cause harm.

Question 7. What is a Safety audit?

Answer: A safety audit is taking time to ensure that the stipulated safety policies are implemented properly. The audit seeks to check whether the safety policies actually

24: Miscellaneous - Questions:

provide the necessary protection from accidents and incidents. Employers and business owners are encouraged to do safety audits regularly to make sure they are not using out-dated safety policies.

Question 8. What is safety tag?

Answer: A safety tag is when a particular safety issue is given priority over the others. The safety tag identifies something that has higher chances of causing accidents especially in workplaces. The tag puts more emphasis on the particular thing so as to provide ample protection.

Question 9. What is safety program?

Answer: A safety program enumerates a number of steps which are put in place to enhance safety. Such a program usually enlightens the people involved on the best ways of enhancing their safety and the safety of the environment.

Question 10. Define an emergency?

Answer: Emergency is a situation which was not foreseen but it has happened as a surprise. These are risky situations which put people at risk but happen when no one is not expecting. They are required to be handled as soon as possible so as to reduce or eliminate the risks.

Question 11. What is work permit system?

Answer: This is a written document which gives permission to undertake a job in a particular area. It is a written document issued by the area in charge to the performer giving authorization to undertake a particular job.

Question 12. What is confined space?

Answer: Confined space is when a worker or workers are working in a place which is enclosed and there is no freedom to move around. There is possibility of heavy and poisonous gases accumulation like H2S (heavier than air). For ex. an area which is small and enclosed or an area where there is only one entry and exit or where a man cannot work comfortably in any location. Working in deep trenches, wells underground places where water pumps are installed etc.

Question 13. What control measures are necessary in confined space?

Answer: Ensure that there is space for air to enter to avoid suffocation. There should be a space for emergency exit in case of situations which might require leaving the space within the shortest time possible.

Question 14. What is LEL?

Answer: Lower Explosive Limit or LEL is the minimum concentration of gases, vapour and dust in air below which flame propagation does not occur on contact with a source of ignition.

Question 15. What is UEL?

Answer: Upper Explosive Limit or UEL is the optimum proportion of gases, vapour and dust in air above which the flame does not occur on contact with a source of ignition.

Question 16. What is excavation?

Answer: This is making a tunnel or a hole by digging on the ground. Excavation can be done by either by a person or a machine depending on the requirements.

Question 17. What is scaffolding?

Answer: It is a temporary platform which is constructed to support workers while working in a construction site. This is especially needed while working on heights where the workers are unable to work from the ground.

Question 18. What is sandblasting?

Answer: Sandblasting is removing unwanted substances such as rust, scales or scales from the old surface using compressed air. The removal of these substances makes the surface becomes as good as new. For ex. the metal plates of Storage tanks are sandblasted to remove rust and other foreign material from them before applying paint.

Question 19. What are the precautions for sandblasting?

Answer: It is important to wear protective clothes while sandblasting to ensure one does not get hurt. We should use the right tools which do not bring about any risk as the substances removed during sandblasting can be harmful.

Question 20. What is personal protective equipment (PPE)?

Answer: This is the equipment that is specifically meant to enhance the safety of individual workers. The equipment is used by individual workers with an objective of enhancing their safety before commencing work or moving in notified zones. Examples are Safety helmets, eye wear, Coverall, Safety Shoes etc.

Question 21. What is work at height?

Answer: Locations where workers are working at considerable height above the ground, where there is a risk of falling. For example, working on storage tanks, cleaning of glass panes of multi-storey offices from outside by using scaffolds etc.

Question 22. What is manual handling?

Answer: When workers do not use any kind of machineries to perform different duties, but they use their hands to carry different jobs especially those which are not too heavy.

Question 23. What are the different types of hazards?

Answer: Some of the most common types of hazards include; mechanical hazards, chemical hazards, electrical hazards, fire hazards, pollution hazards and radiation hazards.

Question 24. What are hazards and injuries in manual handling?

Answer: While doing manual handling a worker can get injured if he is lifting a heavy object. On the other hand, if he is dealing with sharp objects there are a possibility of being cut or pierced by the sharp objects.

Question 25. What are the causes of accidents in manual handling?

Answer: The main cause of accidents in manual handling is failing to follow the stipulated instructions.

Question 26. What is a JSA?

Answer: JSA or Job Safety Analysis is the procedure of analysing job for a specific purpose of making sure that all the hazards are identified.

Question 27. What are advantages of JSA?

Answer: The workers are assured that the working environment is safe. It is also easier to check the various hazards which result in better organization of the projects.

Question 28. What precautions are needed to avoid accidents in manual handling?

Answer: The most important safety measure while doing manual handling is to follow instructions on how to handle different things. To avoid accidents while manual handling it is advisable to always wear the necessary protective clothes. While doing manual handling it is also advisable not to be in unnecessary hurry because this is a major cause of accidents. Working in a hurry makes workers lose concentration and this exposes them to accidents which could have been avoided.

24: Miscellaneous - Questions:

Question 29. Can poor housekeeping be a cause for accidents?

Answer: Yes, it is one of the main reason of accidents at workplace and even at home. It is quite common to observe people tripping on things that might have been left on the floor while they should have been kept somewhere else. Items getting damaged when they come into contact with water or other substances that can easily cause damages. Poor lighting usually causes people to trip on different things.

Question 30. How do you care and maintain hand tools?

Answer: Ensure that the hand tools are properly cleaned after they have been used. Use the hand tools exactly as per as the instruction. Ensure that the hand tools are stored exactly where they should be stored.

Question 31. What precautions are necessary in electrical work?

Answer: When the electricity experiences a problem, the power should be switched off at the main switch to avoid dangers. Always ensure that the person dealing with electrical issues is properly trained and has a license. Never let water near electricity because water is a good conductor of electricity and can cause damages (electrical shock, leading to fatalities). Use high quality wires which are properly coated (insulated) to avoid unnecessary accidents.

Question 32. What are the hazards in construction?

Answer: Falling off from height is one of the main hazards in construction. An object falling on a worker is also a common hazard in construction. Being tripped by objects placed on the ground is also a common construction hazard.

Question 33. What are the general precautions in construction?

Answer: It is important to wear the right protection clothes (PPE) while working in a construction site. Use the tools that are supposed to be especially used for the machineries. Do not leave tools and construction materials unattended on the ground.

Question 34. What are the safety precautions to be observed while working on scaffold?

Answer: Always ensure that the scaffold is strong enough and can support the weight. Wear the right protective clothes especially the ones that are necessary and are supposed to hold one in place while working on the scaffold. Rigger's guidance/presence may be required in working at high-risk scaffold jobs

Question 35. What are the safety rules while using ladders?

Answer: Ensure that the ladder is not slippery by ensuring there are no substances such as oil or water. Ensure the ladder is supported properly and it cannot slip. Make sure that the ladder is strong enough and can support the weight which will be used on it.

Question 36. Can any driver be allowed to operate a fork lift truck?

Answer: A fork lift truck needs to be operated by a Licensed and qualified person who knows how to operate it properly. It should also be positioned in a place where it will not cause any damages and where it has ample space to operate freely. The truck should be removed from the work place when it is not needed anymore.

Question 37. What are the precautions during excavation?

Answer: Ensure that there is enough room for the excavation to be done properly. Use supportive materials to ensure that the tunnel or hole does not collapse.

24: Miscellaneous - Questions:

Question 38. What is a Tool box talk?

Answer: This is the discussion regarding safe work procedures, following of safety rules, job related hazards and the job-related safety aspects. This type of talk plays an important role in ensuring all the safety attributes are in place prior to starting any project.

Question 39. What are the different types of inspections?

Answer: There are 5 different types of inspections and they are; continuous inspection, periodic inspection, intermittent inspection, statutory inspection and special inspection.

Question 40. What is safety management?

Answer: This is the art and science of achieving the safety objectives of the company and therefore come up with the things that should be put in place to ensure that the necessary safety attributes are put in place.

Question 41. What is accident investigation?

Answer: This is trying to get all the facts about the accident that has happened, to know what the causes were and also how to avoid future accidents. This investigation needs to be done as soon as possible after the accident has happened so as to take the necessary precautionary measures.

Question 42. How should an accident be reported?

Answer :To effectively report an accident it is important to include a number of details including; date, time, activity, what happened, the people involved, what did not go right, causes, the suggested corrective actions, the Process Safety Engineer reporting, the project manager and the signatures of all the people who got the report about the accident.

For complete investigation process you can go through the book "Accident & Incident Investigation" available on Amazon:

https://www.amazon.com/dp/B087677K6S

https://www.amazon.in/dp/B086X78YNH

Question 43. What is Management's role in industrial safety?

Answer: The management is responsible for issuing a safety policy for the workers, materials and machines. The safety policy should bring management towards the various aspects in the workplace. Distributing the safety policy to all the interested parties/people involved in the workplace. Arranging for safety inspection which should be done once in every three months and arranging for safety audit which should be done once every year.

Question 44. What are the human factors that usually cause accidents?

Answer: There are a number of human factors that causes accidents and they include; carelessness, hurrying to increase production, laziness, lack of attention, lack of skills, being intoxicated and not using PPE. The Process Safety Engineer might come up with an assortment of policies to enhance safety but it is the responsibility of the workers to ensure that they avoid the various human factors to make sure that there are no accidents.

Chapter 25

CONSTRUCTION RELATED QUESTIONS

Question 1. What is a Safety Program?

Answer:

Safety program can be defined as a method by which accident can be prevented easily by engineering, education, enforcement and enthusiasm.

Question 2. What is Emergency planning?

Answer:

Emergency planning is a control measure by which we can control the accidents, safe guard the people and provide information to media.

Question 3. What is Work Permit System?

Answer:

Work permit system is a "written document" for permission to undertake a job by area in charge or it is written document issued by the area in charge to the performer to undertake the specific job.

Question 3. What is work at height?

Answer:

Any work above 2 meters from ground is called work at height.

Question 4. What Is Excavation?

Answer:

Marking a hole or tunnel by digging the ground by man or machine is called excavation.

Question 5. What is Scaffolding?

Answer:

It is a temporary platform constructed for supporting both men and materials and working safely at a construction site.

Question 6. What is Welding?

Answer:

The process of joining of metals either by electrical process or by gas is called welding.

Question 7. What is Gas Cutting?

Answer:

The process of cutting metals by using oxygen and combustible gas is called gas cutting. The gas cutting job should be carried out very carefully to avoid mishaps due to working at high heat levels of flame.

Question 8. What is Sand Blasting?

Answer:

25: Construction related Questions:

The process of removing rust, dust, dirt, scales and old prints from the old surface using compressed air is called sand blasting.

Question 9. What is LEL?

Answer:

The minimum concentration of vapour, gases and dust in air below which propagation of flame does not occur on contact with a source of ignition is called LEL (Lower Explosive Limit).

Question 10. What is UEL?

Answer:

The maximum proportion of vapour, gases and dust in air above which the propagation of flame does not occur on contact with a source of ignition is called UEL (Upper Explosive Limit).

Question 11. What is Manual Handing?

Answer:

The process of lifting, carrying and stacking materials by men is called manual handing.

Question 12. What is House Keeping?

Answer:

Housekeeping means not only cleanliness but also orderly arrangement of operations, tools, equipment's storage facilities and supplies.

Question 13. What is Personal Protective Equipment?

Answer:

These are equipment used to project the person from hazards such dust, dirt, fumes and sparks etc. It is the barrier between hazard and person.

Question 14. What is Crane?

Answer:

Crane is a tall machine used for moving heavy objects by suspending them from a projecting arm with hook. The cranes can be fixed or movable.

Question 15. What Are the Duties of a Process Safety Engineer?

Answer:

The duties of a Process Safety Engineer involve the following:

Prepare tool box talk

Prepare monthly statistics

Prepare the checklist

Accident reports

Management meetings

Arrange the safety classes/training

Arrange monthly safety bulletin

Inspection of fire extinguisher

Arrange first aid training classes

25: Construction related Questions:

Arrange safety competitions like quiz, slogan, poster competitions exhibition etc.

Question 16. What are the duties of a supervisor?

Answer:

He has to instruct the workers about the work methods and procedures.

He has to maintain discipline among the workers

He has to arrange necessary materials

He has to control quality and cost of the job

He has to supply suitable personal protective equipment to his workers

He should conduct periodical safety meetings.

He should conduct safety inspection of his working area

He should know about the fire fighting equipment

He should be competent to investigate the accident and find out the cause of accident

Question 17. What are the Precautions to be observed in Welding process?

Answer:

Remove all combustion material from the place of welding

Clear the work area and cover wooden floor with fire proof mats. (Welding machine should be kept within the visibility of the welders.

Erect fire resistance screen around the work

All welding cables should be fully insulted

All welding equipment shall be double earthed

Welding area should be dry and free from water

Keep the fire extinguisher / sand ready

Question 18. What are the precautions for Gas Cutting?

Answer:

Keep fire extinguisher nearby

Keep fire watch near by

Remove all combustible material from work area

Use all the necessary PPE

Never put welding gas cylinder inside a confined space

Hoses shall not be laid in path ways

Gas cutting torch should have flash back arrestors

Gas test to be done to check for presence of flammable gas on the work site.

Good housekeeping and ventilation are necessary in working area.

Hose connections should be made properly

Question 19. What are the hazards in Welding?

Answer:

Eye injury

Burn injury from ARC

Electrical shock, Light arc radiation

Heat, light and radiation effect Heat fume

Poisonous gases, Chipped pieces of weld metal hitting the Welder

Fire

Explosion Scattering

Noise Sparking

Sparking

Flying sand

25: Construction related Questions:

Question 20. What precautions are needed to avoid accident in Manhandling?

Answer:

Stand at a safe distance from the load

Sharp edge and bends are removed before lifting a material.

PPE such as safety gloves and safety shoes are to be used.

If the weight is too heavy for one person to lift, then he has to seek assistance.

The pathway is not blocked by obstacles while carrying the load.

The different actions, movements and forces necessary while carrying the load.

Modify the task by using hooks and crow bars.

Mechanical equipment like cranes shall be used.

Modify the objects

Change the way things are used.

Question 21. General safety precautions in Construction?

Answer:

Adequate first aid equipment should be kept ready

Adequate fire fighting equipment should be available

All general electrical rules should be followed

Suitable lighting arrangements should be made for night work

Work men at height should wear safety belts

Work men handling cement should be provided with goggles, rubber gloves and rubber boots & nose mask.

The moving parts of grinding machines used at construction site should be covered with guards

The moving parts of grinding machines used at construction site should be covered with guards

Excavated material should not be kept near the excavated

Very short duration of work red flags must be hoisted and more duration red banners must be stretched

Defective tools should not be used

The worker should not carry tools in his hands when climbing a ladder

Excavation should be guarded by suitable fencing

Chapter 26

SIMPLE CONSTRUCTION SITE SAFETY RULES

Construction sites are dangerous places to work. Every year, thousands of people are injured at work on construction sites

Follow these 10 simple construction site safety rules to keep yourself, and others, safe.

1. Wear your PPE at all times

When you enter the site, make sure you have the PPE you need as it's your last line of defence, should you come into contact with a hazard on site. Safety boots give you grip and protect your feet. Hard hats are easily replaced, but your skull isn't.

It can't protect you if you don't wear it. Wear your hard hat, safety boots and Hi-Viz vest as a minimum, along with any additional PPE required for the task being carried out.

2. Do not start work without an induction training

Each site has its unique hazards and work operations. Make sure you know what is happening so that you can work safely.

Your induction is important. It tells you where to sign in, where to go, what to do, and what to avoid. Don't start work without one.

"Start with your site induction"

3. Keep a tidy site

Construction work is messy, slips and trips might not seem like a major problem compared to other high-risk work happening on the site, but don't be fooled.

Remember to keep your work area tidy throughout your shift to reduce the number of slip and trip hazards. Pay particular attention to areas such as access and escape routes.

4. Do not put yourself or others at risk

One wrong move could put you in trouble. Set a good example, think safe and act safely on site.

You are responsible for your own behaviour. Construction sites are dangerous places to work. Make sure you remain safety aware throughout your shift.

5. Follow safety signs and procedures

Follow the construction safety signs and procedures. The employer should ensure a risk assessment is carried out for your activities.

26: Simple Construction Site Safety Rules:

Make sure you read and understand it. Make sure that the control measures are in place and working before you start.

"Stay safe and follow the rules"

6. Never work in unsafe areas

Make sure your work area is safe. Know what is happening around you. Be aware.

Don't work at height without suitable guard rails or other fall prevention. Don't enter unsupported trenches. Make sure you have safe access. Don't work below crane loads or other dangerous operations.

7. Report defects and near misses

If you notice a problem, don't ignore it; report it to your supervisor immediately. Fill out a "Near-Miss report", an incident report, or simply inform your supervisor.

8. Never tamper with equipment

If something is not working, or doesn't look right, follow rule number 7 and report it. Don't try and force or alter something.

PROCESS SAFETY ENGINEER GUIDE

Never remove guard rails or scaffold ties. Do not remove machine guards. Do not attempt to fix defective

equipment unless you are competent to do so. Do not ever tamper with equipment without authorisation.

"Never remove guard rails"

9. Use the right equipment

Use the correct tool for the job to ensure safe working. Visually check equipment is in good condition and safe to use before you start.

10. If in doubt, ask

Unsure of what to do? Or how to do something safely? Or you think something is wrong? Stop work, and ask. It takes 5 minutes to check, but it might not be so easy to put things right if things go wrong. It's better to be safe than sorry

Chapter 27

PRESSURE RELIEF VALVE: QUESTIONS

Question 1. What is a Pressure Relief Valve?

Answer: Relief valves are automatic valves used on system lines and equipment to prevent over pressurization. Most relief valves simply lift (open) at a pre-set pressure and reset (shut) when pressure drops only slightly below the lifting pressure. System pressure simply acts under the valve disk at the inlet of the valve.

When system pressure exceeds the force exerted by the valve spring, the valve disk lifts off its seat, allowing some of the system fluid to escape through the valve outlet until system pressure is reduced to just below the relief set point of the valve. The spring then reseats the valve.

Question 2. What is the difference between Orifice and Inlet Size?

Answer: The orifice diameter is the internal opening of the valve and is used to calculate the flow capacity of the valve. It's the inside hole. The inlet size is the interface or the size/type of the threads where you attach the valve.

Question 3. How to check whether a Pressure Regulator is leaking or is not working properly?

Answer: The Pressure Regulator leak may be attributed to the following: -

Leak may occur due to debris getting under the diaphragm and preventing it from seating properly and creating gaps which allows air to pass.

This could be caused by having a dirty air environment or by having inadequate filtration.

First check the regulator to ensure that the diaphragm is clean and undamaged.

Make sure that you have an adequate filter upstream and then consider changing the filter element. Dirty air and debris in the system can also cause the diaphragm to tear in which case you would experience the same performance issues.

Question 4. What is set Pressure?

Answer: Liquids tend to be incompressible, meaning they cannot be compressed like air. Liquids can be under

27: Pressure Relief Valve: Questions:

pressure but as soon as the volume changes they immediately lose all pressure (pressure goes to zero).

There are three accepted definition in the industry for liquid applications. They are: start to leak, first steady stream and full flow. For Liquid applications "Set Pressure" is the first steady stream of flow out of the valve.

Question 5. Why do Valves leak?

Answer: It is normal for spring-operated safety valves to exhibit leakage or simmer/warn, as the operating pressure approaches the nameplate set pressure, typically in the 80%-90% range of nameplate set pressure. The ASME Boiler and Pressure Vessel Code do not require a specific seat tightness requirement. A certain level of leakage is allowed per manufacturers' published literature.

Factory Standard Seat Tightness Performance:

Hard Seat Valves – no audible leakage at 20% below nameplate set.

Soft Seat Valves – no audible leakage at 10% below nameplate set.

At very low set pressures (20 psi and below), the ratio of the downward spring force as compared to the upward pressure force is very small. In these cases, it may be impossible to achieve seat tightness. Use soft seat valves for superior seat tightness in applications which fall within the soft seat material temperature limitations.

Question 6. What set pressure should the valve is set to open?

Answer:

Typically, the valve should be nameplate set to open at the MAWP (Maximum Allowable Working Pressure) of the vessel the valve is intended to protect. There is a tolerance to actual set pressure, which means a valve set at 100 psi nameplate may open slightly above or below 100 psi.

Question 7. What is the blow down of a section VIII or Non-code Safety Valve?

Answer: The ASME Boiler and Pressure Vessel Code do not have blown down requirements for Section VIII (or non-code) valves. Blow down may vary from less than 2% to more than 50%, depending on many factors including; valve design, dimensional tolerance variation, where the set pressure falls in the set pressure range of a spring, spring rate/force ratio, warn ring/guide settings, etc. Typical blow down for most valves is 15 % to 30%, but cannot be guaranteed.

Question 8. How does back pressure affect Valve set Pressure and capacity?

Answer: Back pressure reduces set pressure on a one-to-one basis, i.e., a valve set at 100 psig subjected to a backpressure at the outlet of 10 psig will not actuate until system pressure reaches 110 psig. Back pressure drastically reduces capacity; typically, back pressure of 10% of set pressure will decrease capacity by 50%. Specific capacity reduction should be determined by the user on a case-by-case basis by flow testing. Back pressure in excess of 10% of set pressure is not recommended.

Question 9. Does altitude affect set pressure?

Answer: No, gauge pressure (psig) is used to set valves so the effects of weather and altitude on set pressure can be ignored.

Question 10. Why does my Valve actuate/open early?

Answer: It may not be the case, warn/simmer or seat leakage is sometimes mistaken for set pressure. Visible

27: Pressure Relief Valve: Questions:

or audible leakage or system pressure drop is not set pressure. The correct definition of set pressure is as under:

For liquid service- first vertical steady stream

For some valves in air/gas service - first audible sound.

Variance of set pressure is allowed, i.e., a Section VIII air valve with a nameplate of 100 psig set pressure may open from 97 psig to 103 psig, but will be factory set around 102 psig.

Question 11. How high can the system pressure go before the Valve opens?

Answer: Maintain a minimum operating gap of 10% between the system operating pressure and the safety valve nameplate set pressure. Since direct spring-operated safety valves may "Simmer" or "Warn" at 90% of the nameplate set pressure, and since the factory standard leak test performed at 80% of nameplate set pressure, better seat tightness performance can be expected with an operating gap of 20%.

Question 12. Which end should be connected for Vacuum Valves?

Answer: The correct installation often looks backwards from what appears to be correct. A paper instruction tag illustrating the proper connection is attached to each valve. Vacuum valves should have the NPT threads that are cast integral to the body attached to the vacuum source. See the assembly drawing for additional clarification.

Question 13. Why is there a hole in the Valve Body?

Answer: This drain hole is required on some models by the ASME Boiler and Pressure Vessel Code. It is intended to prevent any condensate from accumulating in the body that may freeze or corrode internal valve parts and prevent the valve from opening. The drain hole should be piped away to safely dispose of any discharge or condensate.

Question 14. What mounting orientation should be used to install a Safety Valve?

Answer: Installing a safety valve in any position other than with the spindle vertical and upright may adversely affect performance & lifetime and may not meet code. Installation in any position other than vertical can violate code standards.

Safety and safety-relief valves should be installed vertically with the drain holes open or piped to a convenient location. All piping must be fully supported.

Question 15. How often should the Valve be Tested/inspected?

Answer: Maintenance should be performed on a regular basis. An initial inspection interval of no longer than 12 months is recommended. The user must establish an appropriate inspection interval depending on the service conditions, the condition of the valve and the level of performance desired.

The ASME Boiler and Pressure Vessel Code do not require nor address testing installed valves. The only thing the codes states are design and installation requirements, such as some valves must have a lifting lever. For instance, for section VIII:

"Each pressure relief valve on air, water over 140 degrees Fahrenheit (60 degrees Centigrade), or on steam service shall have a substantial lifting device which when activated will release the seating force on the disk when the pressure relief valve is subjected to a pressure of at least 75% of the set pressure of the valve.

Chapter 28

FREQUENTLY ASKED PERSONAL QUESTIONS

PROCESS SAFETY ENGINEER GUIDE

28: Frequently asked Personal Questions:

Frequently asked Questions

The role of a Process Safety Engineer is to maintain effective work relationships with people from all social backgrounds; he/she should have knowledge of good safety practices.

Question 1. "What would you do in the first week on the job?"

(This question is designed to find out whether you know what you need to know to do the job – whether you know what's important).

Answer:

I would immediately conduct a thorough inspection of the facilities under my responsibility, getting to know the managers and personnel at those facilities, and examining the company's safety record to identify any trends or patterns that need to be addressed quickly.

Question 2. "How would you handle a Plant Manager who thought safety procedures were a waste of time?"

(This question judges an insight into how you might handle opposition and manage potential conflict).

Answer:

I will try to convince the plant manager of the importance of safety using facts and figures (accident claims, down time, lawsuits, etc.), with particular emphasis on how absence through safety negligence would have a detrimental effect on his own team's output.

Question 3. What would you do if someone called and said there had been a serious accident at our facility?

This question is to find out whether you can balance company policy with common sense while keeping a calm and clear head.

Answer:

One of the first things I'd do is familiarize myself with the company procedures ahead of time so I could make sure my response is in alignment. Then, I'd find out what type of first response was needed – both to treat any injured parties and attempt to gain control of and resolve the situation. I would also immediately identify any other agencies or third parties that would need to be notified (such as the ambulance service or the fire brigade) and get them involved quickly."

Question 4. "What would you do if a plant manager asked you to ignore a safety violation?"

This question will test your moral integrity, and see how you balance ethics, professional responsibility, and common sense.

Answer:

"If it were a minor technical violation that was unlikely to result in injury, I might give the manager fair warning to fix it - say, within, 48 hours. After that time, I'd do another inspection, and if the issue still hadn't been addressed, I would take the necessary course of action. Of course, if it were a major hazard, then I'd refuse to ignore it and ensure that any relevant guidelines were subsequently followed."

Question 5. What was the biggest challenge at your last job?

(This is again a common interview question the purpose of which is to get some insight into your personality, as

28: Frequently asked Personal Questions:

well as into what you find difficult. Be careful here, because if you refer to something that's a common occurrence in your prospective role - having to confront people, for instance - then you're unlikely to be selected fort the job).

Answer:

One good answer is to mention something that is likely to annoy your prospective employer, too, such as employees who are always trying to circumvent safety procedures.

Question 6. "How do you handle record-keeping?

(This question helps to determine whether your organizational skills are up to mark, and if you'll place proper importance on protecting the company from liability).

Answer:

"I've found that keeping accurate records is critical in protecting both the company and its employees. If there was an accident, it would be important to be able to prove that we had provided necessary training and had the right policies in place. I also think it's crucial to make a thorough report of an accident, while the details are still fresh. If an injured employee later tells a different story, it's important to have an accurate record."

Question 7. What kind of routine would you think to have in place after 6 months of working here?"

(This question tries to find out two things. Firstly, that you've done your research about the role and you know what the priorities of the day-to-day job are, and secondly, that you're capable of working productively and efficiently with minimum levels of supervision or guidance).

Answer:

A good answer would cover the duties and responsibilities as detailed in the job description, and might touch on things like: how often you'd perform routine inspections, how often you'd conduct training (and what you might focus on), and how you'd keep track of compliance and incident reporting.

Question 8. What is the number-one priority for a Process Safety Engineer?"

(This question tests you whether you and the company share the same values and to identify if you would be a good "fit").

Answer:

You need to ensure that you are aware of your prospective company's values and standards, but at the core of all health and Process safety roles, there are two main priorities: to keep all employees (and any visitors) as safe and risk-averse as possible, and to protect the company from liability in cases of negligence.

Question 9. What is the worst safety violation you've ever seen?

(This question is designed to take stock of your standards. If your "worst violation" is something your prospective employer does every day, you may be out of luck)

Answer:

You should give the most egregious example you can think of – something so bad that it's highly unlikely this company is doing it on a regular basis. Be brief in explaining the details.

28: Frequently asked Personal Questions:

Question 10. Why should we hire you as a Process Safety Engineer?

Answer:

You should emphasize to the interviewer that you like to inspect facilities, processes and people to ensure they follow occupational safety guidelines (OSH). Your commitment to OSH rules should be evident when they talk about their past experiences. Confidence and communication skills will help them enforce the law in cases where others tend to disregard it.

Question 11. Tell me about yourself?
Answer:

Almost every interview will begin with this seemingly simple question. While you may be tempted to provide the interviewer with every detail about your professional and personal life, many aren't looking for such a long-winded answer. Instead, keep your introduction short and to the point. Highlight what you're most proud of, what suits the position best or what makes you right for the job.

Question 12. What made you apply for this Position?

Answer:

A Company doesn't want to hire employees who are just looking for any job. They want to hire individuals who are dedicated to the position, company or industry. To prove you didn't only apply for this position because you applied to every job posting you saw, describe some specific reasons you want the job.

Question 13. What do you know about our Company?
Answer:

Interviewers want to hear that you know a bit about the company you're looking to work for. To prepare for this question, spend some time researching the company, what it does and develops a few talking points that can prove you know the company well.

Question 14. How did you hear about this position?

Answer:

Let the interviewer know whether you've been referred, if you're a fan of the company or if you found the position on a job board. However, if you found the company on a job board, describe the way it stood out.

Question 15. What makes you qualified for this position?

Answer:

This question may be easier to answer depending on your experience. However, knowing what particular experience you have and how it relates to the job can help you answer the question the best way possible.

Question 16. Why should we hire you?

Answer:

If you're directly asked why the company should hire you, don't get intimidated. Be prepared to speak about your accomplishments, skills and abilities. Be ready to sell yourself.

Question 17. What are your biggest weaknesses?

Answer:

This is a bit scary. While you don't want to sabotage your chances of getting the job, you also don't want to give a response too generic that it seems dishonest. Choose an area that you're trying to improve in and explain what you're doing to turn your weakness into strength.

Question 18. What are your biggest strengths?

Answer:

When selecting your strengths to talk about, don't worry too much about what you believe the interviewer wants

28: Frequently asked Personal Questions:

to hear. Instead discuss strengths with confidence and provide clear examples of how you excel in that area.

Question 19. Are you interviewing with any other Companies?
Answer:

While it may seem like a bad idea to talk about the other companies you're interviewing with, letting the company know you're considering other positions can actually work in your favour. Creating the appearance of desire around you and your professional skills can be appealing for the interviewer.

Question 20: What is your ideal work environment?
Answer:

During the interview, the interviewer is also trying to determine how well you will fit with the company. When describing what kind of work environment, you're looking for, be honest about what you need while also staying realistic.

Question 21: Tell me about a time you worked as a Team?
Answer:

Teamwork is crucial for any company to succeed. By asking you this question, the interviewer wants to know you're capable of working with others. Describe a time your team has come together to accomplish a common goal.

Be brief but bring out the salient facts that led to cohesion amongst team members.

Question 22: Why do you want this Job?
Answer:

Getting asked why you want the job can be intimidating. While a better salary, benefits package or location may be the real reason you're looking for the job, you probably don't want to answer that way. Instead, answer this question by talking about qualities of the company or specific roles of the position that make this job the perfect fit for you.

Question 23: When was a time you made a mistake at work?

Answer:

During an interview, you are probably scared of admitting your mistakes. However, mistakes happen. Interviewers know it matters more how you solved the problem. Choose a situation where you made a minor mistake at work and describe what you did to make the situation better.

Question 24. Where do you see yourself in five years?

Answer:

Interviewers want to know that you're looking to progress, especially if you want to move forward within the company. Share what you hope to accomplish in the next few years and how that position and company can help you get there.

Question 25. Tell me about your dream job?

Answer:

Sharing your dream job can help the interviewer understand if this is the right path for you. If your dream job is in a different industry, there is a good chance you'll eventually be leaving the company. When talking about your dream job, relate it back to the position you're applying for.

Question 26. Tell me about your ideal Workday?

Answer:

Explaining your ideal workday can help identify whether or not you'll be happy in the job. If you're looking for a

28: Frequently asked Personal Questions:

schedule or environment that doesn't fit what you're looking for, you probably won't be happy in the job. Be realistic about the day you describe.

Question 27. Why are you leaving your current position?
Answer:

This can be a difficult question to answer. While you don't want to badmouth your current company or manager, you also want to show that the new role is a better fit for you. Focus on what the new position can give you that your current company can't.

Question 28. Tell me about your Management Style
Answer:

If you're applying for a management role, your style may influence how well you fit with the organization. Give an answer that is honest but also fits within the culture of the company.

Question 29. How would your Managers and Co-workers describe you?
Answer:

Try to think up some genuine answers to prepare for this question by pulling from conversations or reviews that you've already had. However, when you answer, remember to be honest in case the interviewer asks your references.

Question 30. What is your most notable Professional Accomplishment?
Answer:

Don't panic if you're asked these questions and you don't have awards or standard accomplishments you can point toward. Speak honestly about something you achieved that truly made you proud.

Question 31. What makes you different from other Applicants?

Answer

While you may not know who the other applicants are, interviewers may ask this question to find out what you think is unique about yourself. Make a list of things you can bring to the table that you think other applicants may not have. Pull from your unique experiences, skills or techniques and relate them to the position.

Question 32. Tell me a time you went above and beyond a Project's requirements?

Answer:

Interviewers don't want to hire someone that only does the bare minimum. Be prepared to explain a time you were asked to do something and you took it to the next level.

Question 33. How do you handle disagreements with your Boss?

Answer:

Disagreements with your boss can happen, but interviewers want to know how you can handle them appropriately and productively. Be ready to talk about your communication skills and problem-solving skills.

Question 34. Where are you in the job-search process?

Answer:

Interviewers want to know what your job-search process has been like. If you've just started applying, you may not actually be prepared to accept a position. Stay honest but let the company know you're searching for the right fit.

28: Frequently asked Personal Questions:

Question 35. What do you do for fun?

Answer:

Your life isn't just about your job. Interviewers want to know that you have hobbies, goals and interests outside of your career. Answer this question honestly, but consider professionalism when you do.

Question 36. Do you have any Leadership experience?

Leadership experience shows interviewers you can take control of a situation when necessary. Whether or not you've had a professional leadership role, discuss a time you led a team or group to accomplish a task.

Question 37. What would you expect out of Management?

Answer:

Your relationship with your manager or supervisor will typically influence how well you do in the company. Knowing what you expect or need out of the managers you work with them determines whether or not you'll be a good fit. This question is another one to answer honestly but realistically.

Question 38. What motivates you?

Answer:

Interviewers want to see that potential employees are driven to accomplish goals and continue moving forward. Knowing what pushes you to wake up every morning and go to work can help them determine if you'll do well with the company. While many of us work for the pay check that comes with it, talk about other motivators like passions, family or interests.

Question 39. What are the biggest challengers you have with your Industry?

Answer:

Interviewers want to know that you recognize your weaknesses and you're looking to change them. For each challenge, also discuss what you're doing to overcome it.

Question 40. What do you hope to accomplish in this position?
Answer:

Discussing what you hope to accomplish in the job shows you've pictured yourself in the company. Relate the specifics of the job description to your professional goals to explain how the position will help you advance your career to the next level in the organization.

Question 41. How do you deal with pressure?
Answer:

We all know jobs can be stressful, so knowing you won't buckle under pressure is important for an interviewer. Talk about some specific things you do to calm your nerves, tackle a situation head on and stay productive.

Question 42. What professional areas would you like to improve?
Answer:

This question is similar to asking about your weaknesses or challenges, but it specifically asks about the areas you hope to grow in. Consider the professional areas you will need to improve in order to advance your career, but also talk about the specific steps you're taking to get there.

28: Frequently asked Personal Questions:

Question 43. What are your expectations for this position?

Answer:

You want your expectations for the position to align with the expectations the company has for you. Use your knowledge of the company, position and job duties to formulate an answer that lays out a few expectations you have. Go through company website and related websites to get an insight for the best answer before the actual interview.

Question44. Would you like to ask us anything?

Answer:

You should have a few questions prepared for the interviewer every time you go into an interview. These questions should relate to the needs of the job, the environment of the job and the expectations of the position.

Ultimately, most of the interview questions asked from the candidates for positions of health and Process Safety Engineers are designed to find out two things: whether you know how to protect people, and whether you know how to protect the company.

As long as your answers highlight your proficiency and commitment to - those two things, then you should do just fine.

Chapter 29

CASE STUDY (CONFINED SPACE)

Case study is one of the tools used in process safety management. The employees can get involved examining the evidence, and understanding why an accident has occurred.

Case studies are interactive and highly effective training method used in the industry.

Workers get to learn how to avoid similar accidents and incidents from the knowledge gained by examining the facts of real work place accidents. This case is about an accident taking place in a confined space.

The Incident

On a summer afternoon, three mechanical technicians arrived at an oil jetty concrete pit where pumps and motors are installed for pumping oil into the ships. The pump was not functioning properly and they had come to repair the pump. When the technicians uncovered the pit, they felt a burning sensation in their eyes.

Technician #1 and Technician #2 planned to go inside the pit whereas Technician #3 would be acting as buddy on top of the pit.

Technician #1 went down into the pit to see what might be causing their eyes to burn. He immediately came out of the pit because it was hot due to warm ambient air. He decided to pour water through water hose into the pit to cool it.

He decided to go down into the pit with water hose along with Technician #2. Both technicians experienced difficulty in breathing and burning sensation in their eyes. Both of them decided to exit the pit because of the intolerable conditions.

Technician #2 climbed out first. As Technician #1 was climbing the ladder to get out, he was overcome by the fumes and fell back into the pit. He became unconscious inside the pit.

Technician #2 went down into the pit in an attempt to rescue the Technician #1, but he was unable to lift Technician #1 and he exited the pit in order to get external help.

Unfortunately, by the time help arrived, employee number one had died of asphyxiation.

The accident investigation revealed that the Technician #1 had tried to extinguish a small cutting torch fire one day before by covering it with sand. Apparently, the fire was not extinguished and smouldered overnight, which resulted in build-up of carbon monoxide gas inside the pit.

After presenting the case, the incident could be analysed by trying to get answers to the following questions. For example:

- What are the potential hazards of confined spaces?
- Were these technicians properly trained and equipped to enter the confined space?

29: Case Study (Confined space):

- Was there a work permit requirement for this confined space? If so, were the technicians familiar with the safety requirements of the permit?
- Was confined space rescue equipment readily available?
- What was the specific hazard in this case that caused the fatality?
- What type of air monitoring should have been done before entering the confined space?

Analysis

a) **Training?** There is no mention on the accident report that the technicians were trained as authorized entrants of confined space. Even if they did receive any confined space entry training, they did not apply what they had learned. The authorized entrants for confined spaces are trained on the hazards of confined spaces, testing procedures, symptoms of lack of oxygen or exposure to toxic gases and chemicals, personal protective equipment (PPE), communication equipment, rescue retrieval equipment, gas mask etc.

b) **Hazard warning?** The technicians entered the space in spite of experiencing burning eyes and unusual heat. Important part of training for confined space workers includes learning about the hazards such as symptoms of lack of oxygen or exposure to gases and toxic chemicals. Workers should never enter a space, and should immediately leave a space, in which they experience signs of hazardous conditions.

c) **Permit requirements?** The confined spaces require a permit before workers can enter the space. The confined spaces have the potential for hazards such as hazardous ambient conditions, heat, combustibility, and entrapment, falls, engulfment etc. By reviewing the permit, the entrants know that they have obtained all the necessary equipment and the atmosphere has been monitored so that they can enter the confined space safely.

d) **Testing?** The Technician #1 died due to lack of oxygen or asphyxiation. If the atmosphere and the pit had been tested prior to entry through a gas tester, this accident would not have occurred. The normal monitoring practices require a check on the oxygen concentration, and check for flammable gases or vapors specially if welding is going to be done in the space and also check for any other toxic chemicals or gases known to be potentially in the space. For spaces in damp conditions or where there is vegetation that is every likelihood of the presence of H2S gas (hydrogen sulphide gas) which is life threatening and is a cause of major facilities.

e) **Rescue procedures and equipment?** The technician #1 who collapsed back into the pit while climbing out could not be rescued because he was not wearing the required rescue equipment. He should have been wearing a full body harness attached to a retrieval line that was connected to a winch type system that could have been used to pull the unconscious worker out of the pit. The other Technician would have had to be trained in confined space rescue procedures.

LIST OF ABBREVIATIONS

An exhaustive list of abbreviations used in Process Safety and other HSE related topics are given below:
(some of them have been used in this book)

ASTM	American Society for Testing Materials
ALARP	As Low As Reasonably Practicable
ASME	American Society of Mechanical Engineers
BPCS	Basis Process Control System
BS	British Standards
CA	Consequence Analysis
CCPS	Centre for Chemical Process Safety
COL	Conservation of Life
CSB	Chemical Safety Board (US)
DHA	Dust Hazards Assessment
ER	Entity Relationship
EER	Energy Efficiency Ratio
EPA	Environment Protection Agency (US)
ERP	Emergency Response Planning
EU	European Union
FEED	Front End Engineering and Design
FMEA	Failure Mode and Effect Analysis
FTA	Fault Tree Analysis
GHS	Globally Harmonized System of Classification & Labelling of Chemicals
HAZID	Hazard Identification
HAZAN	Hazard Analysis
HAZOP	Hazards and Operability Study
HEE	Hazardous Event Evaluation
HF	Human Factors
HIRA	Hazards Identification and Risk Assessment
HSE	Health and Safety Executive (UK)
HSE	Health Safety and Environment
IHA	Intrinsic Hazards Assessment

ISP	Inherently Safer Processes
ITPM	Inspection, Testing and Preventive Maintenance
JD	Job Description
JHA	Job Hazard Analysis
JSA	Job Safety Analysis
KPA	Key Performance Indicators
LOPA	Layer of Protection Analysis
MOC	Management of Change
OD	Operational Discipline
OSHA	Occupational Safety and Health Administration
PDCA	Plan, Do, Check, Act
PHA	Process Hazard Analysis or
PHA	Process Hazard Assessments
PHRA	Process Hazards and Risk Analysis
PPE	Personal Protective Equipment
PRV	Pressure Relief Valve
PSI	Process Safety Information
PSM	Process Safety Management
QRA	Quantitative Risk Assessment
RA	Risk Assessment
RAGAGEP	Recognised and Generally Accepted Good Engineering Practices
RBPS	Risk Based Process Safety
RCA	Root Cause Analysis
SIL	Safe Integrity Level
SIS	Safety Instrumented System
SWMS	Safe Work Method Statement
SWP	Safe Work Practices

ABOUT THE AUTHOR

In the realm of self-growth and safety, P K Singh emerges as a dedicated explorer and advocate, navigating the intricate landscapes of personal development and well-being.

Armed with academic qualifications and experience in the Energy sector, the author is an Electrical Engineer from Delhi University and a certified Trainer from the Institute of Learning & Management, UK. He is a DNV, and Bureau Veritas certified ISO 9001-2015 QMS/EMS Lead Auditor having more than 40 years of experience in handling SHE, Operations, Logistics, HR, Recruitment, and Training functions in various multinational industries in India and abroad.

As an author, he has translated this wealth of knowledge into the written word, offering readers a road-map for personal development and strategies to enhance their safety and well-being. P K Singh's books are characterized by a blend of research-backed insights, practical advice, and a compassionate understanding of the challenges individuals face on their journey toward growth and self-discovery.

Through the pages of his Books, he invites readers to embark on a transformative journey, fostering a deeper connection with the subject for career development.

A
Safety Series Book
from
"pk resources"
India

www.ingramcontent.com/pod-product-compliance
Lightning Source LLC
Chambersburg PA
CBHW052345220526
45465CB00003BA/963